Aeroelasticity of plates and shells

Monographs and textbooks on mechanics of solids and fluids

editor-in-chief: G. Æ Oravas

Mechanics: Dynamical systems

editor: L. Meirovitch

Aeroelasticity of plates and shells

E. H. Dowell

Princeton University
Department of Aerospace and
Mechanical Sciences
Princeton, New Jersey

NOORDHOFF INTERNATIONAL PUBLISHING LEYDEN

A division of A. W. Sijthoff International Publishing Company B.V.,
Leyden, The Netherlands

ISBN 90 286 0404 9

Printed in The Netherlands by
Koninklijke Drukkerij Van de Garde – Zaltbommel

Contents

Preface

This monograph is concerned with a particular aspect of the field of aeroelasticity, namely the aeroelastic behavior of plates and shells. The field of aeroelasticity is inherently of an interdisciplinary nature comprising those physical phenomena which involve significant mutual interaction among aerodynamic, inertial and structural forces. In the standard textbooks* on the subject, it is usually assumed that the phenomena can be adequately understood in terms of linear models and inviscid aerodynamics and that the questions of self-excited stability and response to external disturbances can be readily separated. While these assumptions have been increasingly questioned in recent years with respect to a variety of aeroelastic phenomena, it is in the aeroelastic behavior of plates and shells that these assumptions are invalidated most frequently and where the greatest progress has been made in understanding the effects of nonlinearities, viscous flows, and the complex relation between stability and response. Certain aeroelastic phenomena, e.g., stall flutter and transonic buzz, have been widely recognized to involve significant nonlinearities and viscous effects for many years but nothing that could be termed a systematic theory has been developed to date. In the present monograph we shall present a connected account

---

* Bisplinghoff, R. L., H. Ashley, R. L. Halfman, *Aeroelasticity*, Addison-Wesley, Cambridge, Massachusetts (1955).
Bisplinghoff, R. L. and H. Ashley, *Principles of Aeroelasticity*, John Wiley & Sons, Inc., New York (1962).
Fung, Y. C., *An Introduction to the Theory of Aeroelasticity*, John Wiley & Sons, Inc., New York (1958).
Scanlan, R. H. and R. Rosenbaum, *Introduction to the Study of Aircraft Vibration and Flutter*, The MacMillan Company, New York (1951).

of the aeroelasticity of plates and shells; Chapter 1 discusses the physical nature of the problem, experimental methods, and the basic ingredients required to understand and model the phenomena; Chapters 2, 3 and 4 discuss the theoretical foundations and methods of analysis treating, in turn, linear aeroelastic models, their nonlinear counterparts and the requisite aerodynamic theory; Chapter 5 discusses the relationship between self-excited and externally excited response; Chapter 6 discusses experimental results including correlation with theory; Chapter 7 suggests future directions of research for achieving a better understanding of the aeroelasticity of plates and shells as well as other aeroelastic phenomena where the methods discussed herein may be applicable. The implications of the results for design practice have been thoroughly treated elsewhere* and will not be considered in this monograph.

The present volume should be suitable as a reference source for the practicing aeroelastician or as a text for a graduate seminar in modern topics of aeroelasticity.

Although most of the work in the field has been motivated by applications to aerospace engineering, other fields of technology have generated similar problems which may be effectively attacked by similar methods. In particular the acoustician and structural dynamicist concerned with sound-structural interaction will find the present volume in many respect complementary to the book by Junger and Feit**.

The author would like to acknowledge the benefit of many fruitful discussions with Professor John Dugundji of the Massachusetts Institute of Technology, Dr. H. M. Voss of the Boeing Company, and Dr. C. Samuel Ventres of Bolt, Beranek and Newman, over a period of several years. He has, of course, greatly benefited from the contributions by various authors to our present understanding of the aeroelasticity of plates and shells. Many of these are individually cited in the text. He would also like to thank his colleague, Dr. Francis C. Moon, for several valuable suggestions regarding Chapter 1 and Mrs. Linda Ludwig, Mrs. Elfriede Takats and Mr. Anthony Poli for their help in the preparation of the manuscript.

---

* Panel Flutter, NASA Space Vehicle Design Criteria, *NASA SP-8004* (1972).

** Junger, M. C. and D. Feit, *Sound, Structures, and Their Interaction,* The M.I.T. Press, Cambridge, Massachusetts (1972).

Nomenclature

A	area
a	plate length
a_i, a_r	generalized coordinates
b	plate width
b_k	generalized coordinates
D, D_x, D_y, D_1, D_{xy}	plate stiffnesses
d	cavity depth
E	modulus of elasticity
F	Airy stress function
F_m	acoustical normal mode function
G	shear modulus, $\dfrac{E}{2(1+\nu)}$
H	maximum rise height of curved plate
h	plate thickness
$K = \omega\left(\dfrac{ma^4}{D}\right)^{\frac{1}{2}}$	non-dimensional frequency; also spring stiffnesses
$L = T - U$	Lagrangian
M	Mach number
M_n	generalized mass
$m = \rho_m h$	panel mass/area
N_x, N_y, N_{xy}	in-plane stress resultants
Δp	(aerodynamic) pressure loading on panel
p	fluid pressure
$P = \dfrac{\Delta p a^4}{Dh}$	non-dimensional pressure

Q_n	generalized (aerodynamic) force
$q = \dfrac{\rho_\infty U_\infty^2}{2}$	dynamic pressure *
q_n	generalized coordinate
R	radius of cylinder; also gas constant; also Reynolds number
R_x, R_y, R_T	non-dimensional in-plane loads
T	kinetic energy; also temperature
ΔT	temperature differential
t	time
U	fluid velocity; also strain energy
V	volume
u, v, w	plate displacements; also fluid velocities
W	work
x, y, z	spatial coordinates
$z_p(x, y)$	initial shape of curved panel
α	thermal expansion coefficient
$\beta = (M^2 - 1)$	
β_n	modal structural damping coefficient
γ	ratio of specific heats for a gas
δ	boundary layer thickness; also virtual operator
δ_{mn}	Kronecker delta
ε	strain
ζ	percent of critical structural damping
$\eta = y/b$	
λ_s	Lagrange multipliers
$\lambda^* = 2qa^3/D$	non-dimensional dynamic pressure
$\lambda = \lambda^*/\beta$ or λ^*/M	
$\mu = \rho_\infty a/m$	non-dimensional mass ratio
ν	Poisson's ratio
$\xi = x/a$	
ρ	fluid density
ρ_m	panel material density
σ	stress

* The name is somewhat of a misnomer; q does not vary time for the problems of interest here.

$\tau = t\left(\dfrac{D}{ma^4}\right)^{\frac{1}{2}}$	non-dimensional time
ϕ	velocity potential
ψ	stream function
ψ_m, ϕ_r	structural modes; also fluid modes
ω	frequency

Superscripts

c	cavity
E	external (to panel)
M	motion dependent
S	static
$-$	amplitude of complex number; also mean property
$\wedge$	perturbation
$'$	$\dfrac{\partial}{\partial x}$
$\cdot$	$\dfrac{\partial}{\partial t}$

Subscripts

c	cavity; also convection
D	divergence
d	dynamic
F	flutter
f	flexible
I	imaginary
p	peak, i.e. maximum of a sinusoidal oscillation ($\xi = .75$, $\eta = .5$ unless otherwise noted); also panel
R	rigid; also real
r	random
rms	root mean square
s	support
w	wall
∞	mean flow conditions outside fluid boundary layer

1

Physical nature of the problem and experiments

The aeroelasticity of plates and shells is a subject which has been intensely studied for the past two decades. The studies have ranged from *ad hoc* investigations to solve an immediate design problem to research studies directed toward obtaining a fundamental understanding of the various physical phenomena. Structural failures have occurred as a consequence of aeroelastically induced motion on a variety of high performance operational and research aircraft, launch vehicles for spacecraft, and jet engines. When the problem first arose it was widely assumed that it was an interesting but not fundamentally different aspect of the general field of aeroelasticity. However it quickly became clear that in at least one respect the problem was substantially different from earlier aeroelastic problems. Using the standard methods for aeroelastic analysis one frequently found the theoretical data to be in substantial disagreement with the corresponding experimental data. This led to many of the subsequent studies which attempted to resolve the theoretical-experimental discrepancies. A considerable source of difficulty was that many of the early experimental data were obtained using (by present standards) rather inaccurate techniques and uncontrolled test conditions.

In recent years most of these early difficulties have been overcome. Careful, controlled experiments have been performed and substantial advances in the theory have been made. Most importantly the data from theory and experiment have been demonstrated to be in satisfactory agreement for a considerable range of parameters. Based upon these more recent results, however, it has become clear that there are several aspects of the aeroelasticity of plates and shells which are not adequately described by the usual aeroelastic models. Among these are structural nonlinearities and viscous flow effects. Because of structural nonlinearities the dynamic aeroelastic instability of plates and shells ('panel flutter') does not usually lead to immediate catastrophic failure, as is often the case for lifting surfaces, but rather is frequently a sustained constant amplitude oscillation which may lead to long time fatigue failure. Moreover, viscous shear flow effects can be important when the shear or boundary layer thickness is comparable to the plate length or even substantially less, particularly in the transonic-low supersonic flow regime. The thin fluid boundary layer region next to the panel can significantly modify the effective Mach number and dynamic pressure of the flow.

Finally, it has become clear that the relationship of forced panel response (due to excitation by 'external' boundary layer pressure fluctuations) to self-excited response, 'flutter', must be reconsidered. Considering the two separately, as in the case of 'gust response' (structural response due to forced excitation by atmospheric turbulence) and 'flutter' of lifting surfaces, is often inadequate for plates and shells because of the limited amplitude nature of the flutter response and the difficulty of distinguishing between forced and self-excited response.

Based upon these principal themes, we shall discuss the present state of knowledge of the aeroelasticity of plates and shells. The discussion is organized to give a logical presentation of ideas and no attempt is made to give a chronological discussion. Several comprehensive historical surveys of the literature are available [1–9]. Also the applications of the methods and results in design practice is not extensively treated here, as it is adequately treated elsewhere, and again the reader is referred to the available literature [8–9].

We shall first turn our attention to the experimental evidence, and methods for obtaining such, which shall provide a basis for the formulation of the subsequent theoretical models.

1.1 Physical nature of the problem

Let us begin our discussion by considering a wind tunnel experiment for a panel at supersonic speeds. See Fig. 1.1 for typical plate and shell geometries. Normally the test would begin with the airflow over the panel

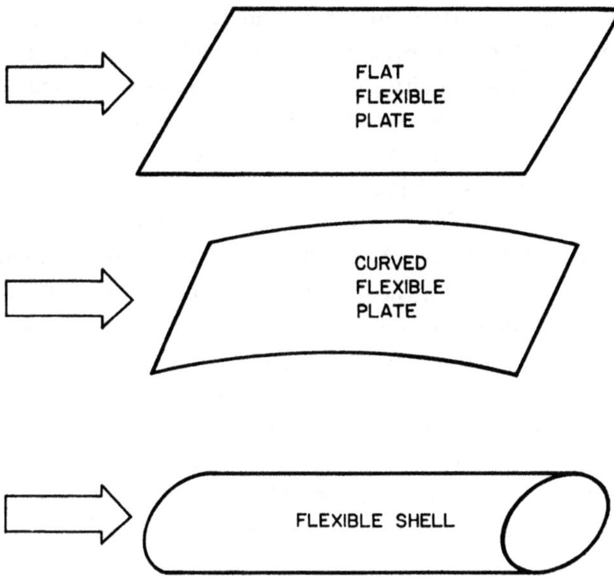

Fig. 1.1 Flow over plate and shell geometries.

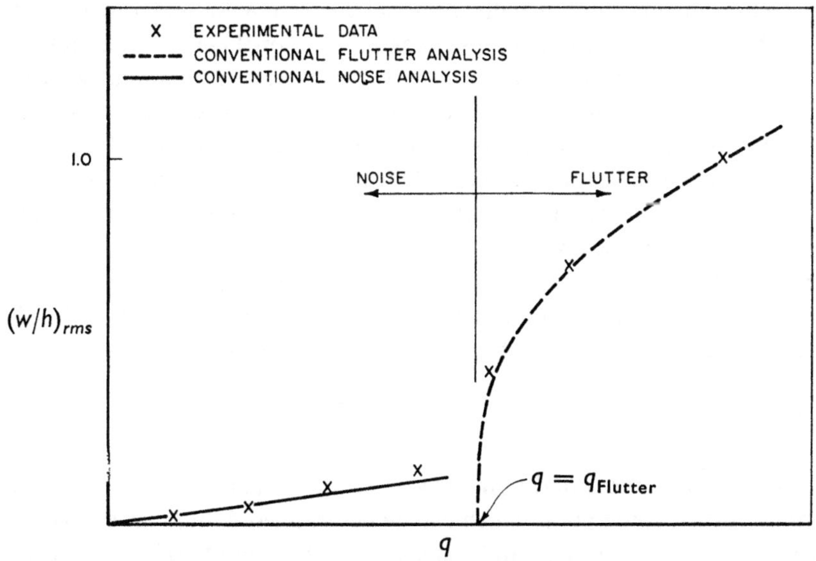

Fig. 1.2 Schematic of panel response.

established at some fixed Mach number. The flutter boundary is sought by increasing the fluid dynamic pressure, q, (one-half the product of density times the square of flow velocity), with Mach number held fixed. For small q we are below the flutter boundary and random oscillations of the panel are observed with dominant frequency components near the lower panel natural frequencies; the panel is responding to pressure fluctuations in the turbulent boundary layer. The maximum oscillation amplitude is some small fraction of the panel thickness. As the flutter boundary is exceeded at some critical dynamic pressure, q_{flutter}, the oscillation becomes nearly sinusoidal with an amplitude which grows to on the order of a plate thickness when q is increased from 25–100% beyond the value at which flutter begins. See Fig. 1.2 for a *schematic* of this behavior. This figure is a freely drawn representation of results from several experimental programs [10–16]. As $q \to q_{\text{flutter}}$, the panel motion itself creates significant pressure fluctuations which in turn modify the panel motion. This 'aeroelastic feedback' mechanism is the fundamental source of the dynamic instability known as flutter.

The schematic nature of Fig. 1.2 should be emphasized. No flutter experiment has yet been performed over the entire range of dynamic pressure indicated. Most experiments have focused on determining q_{flutter} only and hence measurements have usually been taken over a limited range of q. One of the major limitations of the present state-of-the-art is the lack of data over an extensive range of q at supersonic speeds. A preliminary study has been made for subsonic flow to obtain data of this type where the aeroelastic instability is static, 'divergence', rather than dynamic [17]. This 'divergence' is normally of less technological importance than 'flutter'. A very recent study at supersonic speeds provides an initial effort in this direction for 'flutter' motion [18].

We shall defer a specific discussion of structural response to turbulent boundary layer pressure fluctuations *per se* until Chapter 5 and shall initially focus our concern on the flutter behavior of plates and shells.

Identification of onset of flutter

The experimentally measured point at which flutter begins is more a matter of definition than some point which can be determined with great precision. Typically the onset of flutter can be specified within about 10% in terms of dynamic pressure. There has been some effort to obtain a more precise experimental determination of the flutter boundary via admittance techniques and the concept of a linear plate impedance [19].

Character of post-flutter oscillation

The post-flutter behavior is largely dominated by the nonlinearities of the system. Most prominent of these is the nonlinear structural coupling between bending and stretching of the plate. As the plate bends it also stretches thereby inducing a tension in the plate. The post-flutter oscillation represents a balance between the (unstable) linear plate and fluid forces and this tension which increases the effective plate stiffness. Recognizing this balance one can obtain qualitative estimates of the flutter amplitude by order of magnitude considerations [20].

Failure due to flutter

Not a great deal of study has been directed toward failure mechanisms. However, at least two are readily identifiable and have occurred in practice. If the stress amplitude due to flutter exceeds the yield stress of the plate material over a substantial portion of the structure then catastrophic or rapid failure occurs; on the other hand, if the stress is relatively small then fatigue or long-time failure may occur. From a knowledge of stress amplitude and frequency of the oscillation, an estimate of the fatigue life may be made [21].

1.2 Experimental methods

In an actual experiment, much care and attention to detail are required. The essential elements of conducting an experiment are discussed here.

Pre-wind tunnel preparation

Models and vibration tests: The success of a wind-tunnel flutter test is largely determined before the test specimen is placed in the tunnel. The first task, and often the most difficult, is the construction of the panel models and their boundary support. Because of the dynamic pressure ranges available in most supersonic wind tunnels the plate thickness/length ratio must typically be quite small, 10^{-2}–10^{-3}, in order to obtain flutter data over a wide range of other structural and aerodynamic parameters. This becomes a particular problem at high supersonic Mach number where the dynamic pressure required for flutter increases. The fabrication of these thin plates and the mounting on their boundary supports without inducing significant stresses is a major concern. At lower Mach numbers where thicker plates may be used this is a lesser problem. Experience has shown that resonant vibration tests to determine panel natural frequencies and modes have proven excellent indicators of the quality of model construction. If a good job has been done, the measured frequencies and modes will agree closely with their theoretical counterparts, within say 10%. Conventional practice is to use acoustic excitation for vibration tests and strain gauges and capacitance or inductance deflection transducers to measure structural responses.

Pressure differential and thermal stresses: The next step is normally to determine the sensitivity of the models to environmental factors, particularly static pressure loadings and thermal stresses due to a temperature differential between model and support. These effects may require considerable effort to control and measure in the wind tunnel with the thinner plates being particularly delicate. Hence, it is essential that the sensitivity of the model to these factors be determined early in the experimental program by resonant vibration tests. The sensitivity can also be estimated by theoretical means but vibration tests are generally more informative particularly with respect to assessing in-plane edge support conditions and not all that difficult to perform.

Cavity effect: Another concern may be the dynamics of the air in an enclosed cavity beneath the plate. If the cavity depth is on the order of or greater than the plate length or width, the air in the cavity will not play an important role (as far as plate motion is concerned). However, for shallow cavity depths, which are sometimes dictated by practical considerations such as wind tunnel blockage, vibration tests with and without the cavity will be desirable as well. Actually, the resonant vibration test will show a much larger cavity effect than will be present in the wind tunnel since the ambient density and pressure are much larger than normal tunnel conditions. The cavity effect is not always negligible in the tunnel, however.

Wind tunnel flutter testing

Tunnel and instrumentation selection: After the pretunnel program has been completed it remains to mount the panel and its support in the wind tunnel and to install the appropriate instrumentation with supporting equipment. One might also say that it remains to choose an appropriate wind tunnel; however, in practice the choice is often rather limited. Generally speaking, one wants a wind tunnel with the appropriate Mach number range (generally supersonic), largest possible size and dynamic pressure range, and with good control over tunnel temperature.

With regard to tunnel mounting, two techniques have been used: (1) flush wall mount and (2) mid-tunnel splitter plate of an appropriate aerodynamic shape. Generally, the former is more desirable from the point of view of easier access though practical considerations may preclude its use in some tunnels.

The choice of instrumentation is usually made so that the temperature of the plate model and support, pressure differential across the plate, plate strain, deflection and frequency can be measured. This leads to the use of thermocouples, pressure (0.01–1.0 psi) transducers and strain gauges. Capacitance or inductance deflection transducers are often used though other types have been employed successfully. Lightweight, if possible noncontacting, transducers are used which will not alter the dynamic characteristics of the plate nor disturb the aerodynamic flow.

Flutter testing techniques: Once the model is installed most of the work is done and the test is normally carried out by increasing stagnation pressure (proportional to dynamic pressure) at fixed Mach number and stagnation temperature. Achieving flutter by increasing the dynamic pressure is not the only experimental technique which may be employed, however. For example, Lock and Fung [22] have varied Mach number at constant stagnation pressure. On the other hand, the Langley group in their many tests of the flutter of buckled plates (see e.g., [23]) have usually penetrated the flutter regime by heating the panel at constant stagnation pressure to induce compressive thermal stresses which decrease plate stiffness. The latter investigations have employed a short duration 'blowdown' wind tunnel where changing stagnation pressure in precise steps is impractical. Although the use of a short duration tunnel is feasible, it imposes severe requirements with regard to the determination and control of the plate environment, particularly with regard to thermal stresses and static pressure loading. Generally the use of a continuous flow wind tunnel is preferred so that at each value of stagnation pressure and Mach number, sufficient time is available for the temperature and static pressure loading to be brought to the desired equilibrium values.

Scaling and similarity laws: When extrapolating data to new panel designs or planning an experimental program it is important to consider the appropriate scaling laws. Because it is frequently possible to place a geometrically full scale panel in a wind tunnel for testing, there is sometimes an unfortunate and unfounded belief that complete full scale simulation has been achieved. Rarely is this the case. Usually thermal stresses, pressurization loads, flow density, boundary layer and/or other parameters will not be correctly scaled. Hence one must rely upon theory to assess these variables or conversely sacrifice the full scale geometry to help scale correctly other variables. A valuable general discussion of the problems involved which is applicable to panel flutter is given in [24]. If some parameter is thought to be well understood by theory it may be wise to sacrifice proper scaling of that parameter in order to obtain a more nearly true scaling for some other parameter which is not as well understood. See Chapter 6 for a more extended discussion of these matters.

Flight flutter testing

Many of the experimental techniques are, of course, similar to those of wind tunnel flutter testing. In particular, for pretunnel in the previous discussion one may insert preflight test. The reader may consult [25] and [26] wherein the flight test results of the X-15 (hypersonic research aircraft) and S-IVB (launch vehicle for Apollo moon exploration program) panels are discussed.

Because one can frequently mount full scale panels in a wind tunnel and also as a consequence of panel flutter frequently not resulting in major structural failure, there has been a tendency to use wind tunnel tests for proof of adequate design to prevent flutter rather than flight tests. This is in contrast to the situation with regard to flutter of major lifting surfaces where normally flight test programs are undertaken.

1.2 Experimental methods

Role of experimental methods

Vibration testing to determine panel modes and frequencies is very valuable for assessing the adequacy of the structural theory used and is essential for new panel configurations for which no previous experience exists. It frequently provides necessary information for evaluating the theoretical model as far as structural stiffness is concerned. Wind tunnel flutter testing is, of course, also very valuable, but unfortunately very expensive. For designs which are significantly affected by a large number of parameters, it may be impossible to scale all of them properly in a given wind tunnel facility. Hence, partially scaled tests may be required, relying on theory to assess the unscaled parameters. Flutter testing is presently considered essential to evaluate aerodynamic boundary layer effects and to determine amplitudes, stresses and frequency of the flutter oscillation although theoretical analyses are improving rapidly. (See Chapter 6 for further discussion of the above points.)

High Mach number flows are the most difficult to handle experimentally because of the extreme temperature environment and low dynamic pressures available in most wind tunnels. Early experimental flutter data (*pre* 1963) are generally unreliable and even today great care must be exercised to control and measure accurately all significant parameters in a flutter test. The best available experimental methods have given satisfactory results and may be used with confidence. Such methods are, unfortunately, costly and thus for economic reasons, if none other, theoretical methods are important.

We shall defer a discussion of quantitative experimental results until we can compare them to analogous theoretical results. Now we shall turn to the construction of appropriate theoretical models to obtain such results based upon this introductory discussion of the physical phenomena of interest.

2

Linear theoretical aeroelastic models

2.1 Illustrative model

We shall first consider linear models which, with a minimum of complexity, illustrate the essence of some of the physical phenomena which can be described by theoretical means. For definiteness we shall treat the flutter of a one-dimensional ($a/b = 0$), flat plate with one side exposed to a fluid flow. See Fig. 2.1. The specific nature of the aerodynamic theory will be left open and we shall deal with various types of aerodynamic forces in general terms.

Structural and aerodynamic operators

The equation of equilibrium for the plate is:

$$D \frac{\partial^4 w}{\partial x^4} + m \frac{\partial^2 w}{\partial t^2} + \Delta p = 0 \tag{1}$$

where

- w – plate transverse displacement
- D – plate bending stiffness
- m – plate mass/area
- x – spatial coordinate in the direction of the fluid flow
- t – time
- Δp – aerodynamic pressure, positive in direction opposite to w

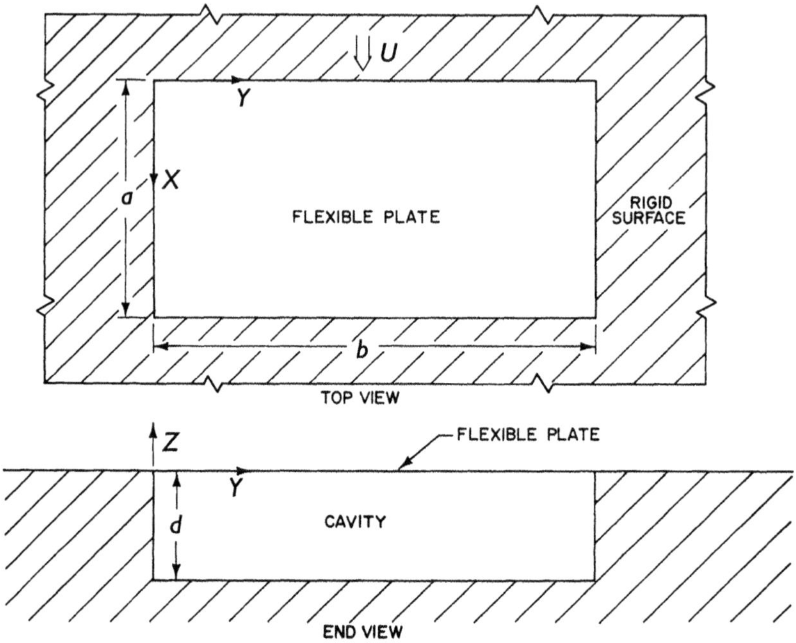

Fig. 2.1 Panel geometry.

Normally the aerodynamic pressure may be considered as the sum of two parts, one given by the pressure fluctuations on the plate in the absence of any plate motion, e.g., due to turbulent boundary layer fluctuations, and the other due to the plate motion itself. The superposition of these two contributions to form the total aerodynamic pressure implies that the plate motion and the consequent portion of the aerodynamic pressure are sufficiently small so that the turbulent pressure fluctuations themselves are not substantially modified. Hence

$$\Delta p = \Delta p^M + \Delta p^E \tag{2}$$

where

Δp^M – aerodynamic pressure due to plate motion
Δp^E – 'external' aerodynamic pressure independent of plate motion, e.g. turbulent pressure fluctuations

Δp^E is taken as prescribed from experiment or separate analysis while Δp^M may be related to w in general (see Section 4 for precise restrictions) by an expression of the form

$$\Delta p^M(x, t) = \rho_\infty U_\infty^2 \left\{ \frac{1}{M}\left[\frac{\partial w}{\partial x} + \frac{1}{U_\infty}\frac{\partial w}{\partial t}\right] + \int_0^{+}\int_0^a A(x-\xi, t-\tau)\left[\frac{\partial w}{\partial \xi} + \frac{1}{U_\infty}\frac{\partial w}{\partial \tau}\right] d\xi\, d\tau \right\} \tag{3}$$

A is a moderately complicated function which depends parametrically on Mach number, $M \equiv U_\infty/a_\infty$. An appropriate non-dimensional aerodynamic time scale is tU_∞/a where a is the plate length. For harmonic motion the corresponding frequency scaling is $\omega a/U_\infty$. A very popular assumption, for obvious reasons, is to simplify Equation (3) by dropping the second (integral) term. We shall explore the consequences of and justification for this simplification later. Note that the integral term describes the effects of spatial and temporal memory, i.e., the pressure at a particular point, x at a particular time, t, is influenced by the motion at all points, $0 < \xi < a$, at all previous times, $0 < \tau < t$. The neglect of the memory effect leads to the so-called 'piston theory' approximation [3]

$$\Delta p^M = \frac{\rho_\infty U_\infty^2}{M}\left[\frac{\partial w}{\partial x} + \frac{1}{U_\infty}\frac{\partial w}{\partial t}\right] = \rho_\infty a_\infty \left[U_\infty \frac{\partial w}{\partial x} + \frac{\partial w}{\partial t}\right] \tag{4}$$

This is recognized as the relation between the motion of and pressure on a piston in a tube where the total piston velocity includes both a convection term, $U_\infty(\partial w/\partial x)$, as well as the direct velocity, $\partial w/\partial t$. Here the plate is the equivalent piston and the fluid 'tube' is perpendicular to the plate.

Modal expansion methods and reduction to discrete coordinates

The system of equations, (1), (2) and (3), comprises our (linear) theoretical model. We may determine solutions in a variety of ways. For our purpose here an expansion of w in terms of its structural normal modes, ψ_m, will be most informative. The ψ_m are the eigenfunctions and ω_m the natural frequencies or eigenvalues of Equation (1) for $\Delta p \equiv 0$. That is, the ω_m and ψ_m satisfy

$$D\frac{d^4\psi_m}{dx^4} - m\omega_m^2\psi_m = 0 \tag{5}$$

The ψ_m must also satisfy the same boundary conditions as w and consequently an orthogonality condition holds with respect to mass, m,

$$\begin{aligned}\int m\psi_m\psi_n dx &= M_n \text{ for } m = n \\ &= 0 \quad \text{for } m \neq n\end{aligned} \tag{6}$$

and also with respect to stiffness, D,

$$\begin{aligned}\int D\frac{d^2\psi_m}{dx^2}\frac{d^2\psi_n}{dx^2}dx &= \omega_n^2 M_n \text{ for } m = n \\ &= 0 \qquad \text{for } m \neq n\end{aligned} \tag{7}$$

Both (6) and (7) follow from (5) and associated boundary conditions, e.g., see any standard text on structural vibrations.

Expanding w we have

$$w(x, t) = \sum_{m=1}^{\infty} q_m(t)\psi_m(x) \tag{8}$$

where of course the q_m are to be determined. Introducing (8) into (1) gives

$$\sum_m D q_m \frac{d^4\psi_m}{dx^4} + m\frac{d^2q_m}{dt^2}\psi_m + \Delta p = 0 \tag{9}$$

We obtain a system of equations for the q_m by employing the averaging procedure known as Galerkin's method. Multiplying (9) by ψ_n, say, and integrating over the plate we obtain

$$M_n\left[\omega_n^2 q_n + \frac{d^2q_n}{dt^2}\right] + \rho_\infty U_\infty^2 Q_n = 0 \qquad n = 1, \ldots, \infty \tag{10}$$

where

$$Q_n \equiv \int \frac{\Delta p}{\rho_\infty U_\infty^2} \psi_n \mathrm{d}x \tag{11}$$

In evaluating the first term of (10) we have used (7) and an integration by parts. If $Q_n \equiv 0$ in (10) we have an uncoupled system of simple harmonic oscillators which is a consequence and benefit of expanding w in terms of its normal modes. Now let us evaluate Q_n by substituting (8) into (3) and the result into (11). Carrying out the several integrations we obtain

$$Q_n = \sum_m Q_{mn}^M + Q_n^E \tag{12}$$

where

$$Q_{mn}^M \equiv \frac{1}{M}\left[q_m \int_0^a \psi_m' \psi_n \mathrm{d}x + \frac{\dot{q}_m}{U_\infty} \int_0^a \psi_m \psi_n \mathrm{d}x\right] + \int_0^t \left[q_m(\tau) A_{mn}^q(t-\tau) + \frac{\dot{q}_m}{U_\infty}(\tau) A_{mn}^{\dot{q}}(t-\tau)\right] \mathrm{d}\tau \tag{13}$$

and

$$\begin{aligned} A_{mn}^q &\equiv \int\int_0^a A(x-\xi, t-\tau)\psi_m'(\xi)\psi_n(x)\,\mathrm{d}\xi\,\mathrm{d}x \\ A_{mn}^{\dot{q}} &\equiv \int\int_0^a A(x-\xi, t-\tau)\psi_m(\xi)\psi_n(x)\,\mathrm{d}\xi\,\mathrm{d}x \end{aligned} \tag{14}$$

and

$$Q_n^E \equiv \int_0^a \frac{\Delta p^E}{\rho_\infty U_\infty^2} \psi_n(x) \mathrm{d}x \tag{15}$$

Equation (10), observing (12) *et. seq.*, is a system of coupled integral-differential equations in time for the modal amplitude q_m. The coupling is entirely aerodynamic as a result of using structural natural modes. (We remark that because all structural boundary conditions are satisfied by the natural modes the above equations could also have been obtained via Hamilton's principle and the Rayleigh-Ritz method.) For a prescribed external pressure field, Δp^E, the plate response as characterized by the q_m may be determined as a function of time. For example, this may be accomplished by step-by-step numerical integration on a digital computer.

Dynamics of discrete aeroelastic system

The external pressure fields of interest are normally random, see Fig. 2.2. The plate response will also as a consequence be random, see Fig. 2.3. However, the plate will act as a mechanical filter and respond primarily to those frequencies of the external pressure field near the structural natural frequencies. Moreover as the flow dynamic pressure increases above a critical value the plate response will change from random to highly ordered with increasing amplitude. This latter behavior signals the onset of a dynamic aeroelastic instability, 'flutter'. In the linear model any disturbance to the panel results in an exponentially growing oscillation with increasing time. In many physical situations structural nonlinearities come into play at sufficiently large deflections and restrain the plate to finite amplitude, resulting in the so-called 'limit cycle oscillations'. We shall discuss this nonlinear behavior subsequently in Chapter 3.

As an alternative to the time domain analysis one may treat the problem in the frequency domain. For approximate analytical solutions the frequency domain may be more advantageous. Taking the Fourier Transform of (10)–(14), or simply assuming simple harmonic time dependence for all quantities, e.g.,

$$w = \bar{w}(x)e^{i\omega t}$$

we have from (10)

$$M_n[\omega_n^2 - \omega^2]\bar{q}_n + \rho_\infty U_\infty^2\{\sum_m \bar{Q}_{mn}^M + \bar{Q}_n^E\} = 0 \tag{16}$$

Now $\bar{Q}_{mn}^M$ may be written

$$\bar{Q}_{mn}^M \equiv \bar{\bar{Q}}_{mn}^M \bar{q}_m \tag{17}$$

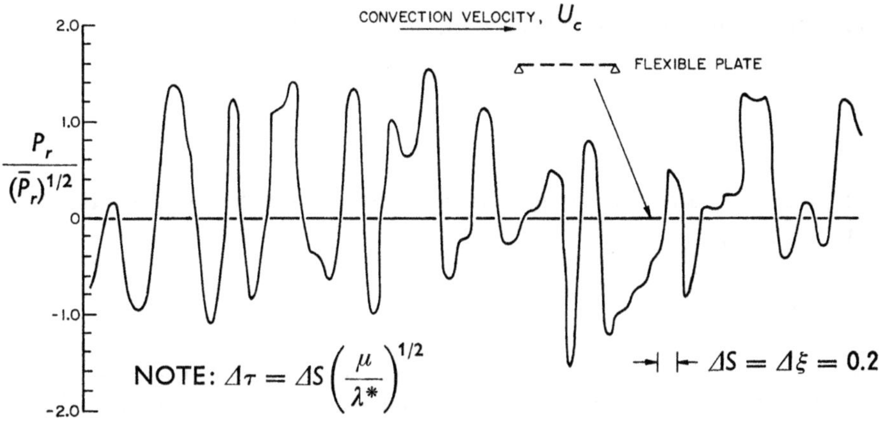

Fig. 2.2 Random external pressure field.

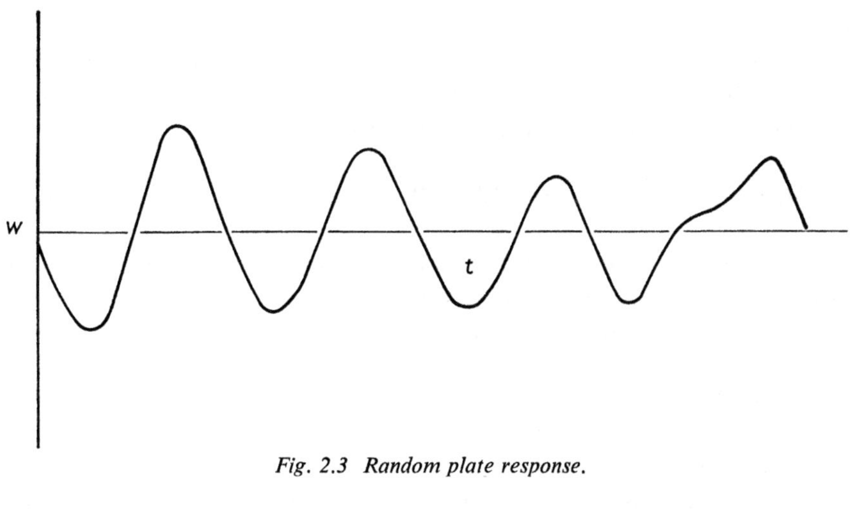

Fig. 2.3 Random plate response.

where the aerodynamic coupling coefficient $\bar{Q}^M_{mn}$ is defined by (17) and is given in detail by taking the Fourier Transform of (13) and identifying the coefficient of $\bar{q}_m$. For a specified $\bar{Q}^E_n$ we may formally solve (16) for the plate model amplitudes by matrix inversion if we truncate to say N modes. In matrix notation,

$$\{\bar{q}\} = \rho_\infty U_\infty^2[M_n[\omega^2 - \omega_n^2]\delta_{mn} - \rho_\infty U_\infty^2 \bar{\bar{Q}}^M_{mn}]^{-1}\{\bar{Q}^E\} \tag{18}$$

Defining a matrix of transfer functions with elements

$$H_{ij} \equiv \frac{\partial \bar{q}_i}{\partial \bar{Q}_j^E} \tag{19}$$

the inversion need only be performed once and the response to any Q_n^E determined by

$$\{\bar{q}\} = [H]\{\bar{Q}^E\} \tag{20}$$

Using the concept of power spectral density the response to a random input may be obtained using the frequency domain description. Formally there is no difference from the usual case discussed in standard texts on random structural vibrations [27]. However, the transfer function is fundamentally more complicated as a result of including the aerodynamic motion forces, i.e., we have an aeroelastic rather than a purely mechanical filter.

Of course under certain circumstances the transfer function may not exist, i.e., the inversion indicated in (18) cannot be performed. In such cases the determinant of coefficients will be zero.

$$|M_n[\omega^2 - \omega_n^2]\delta_{mn} - \rho_\infty U_\infty^2 Q_{mn}^M| = 0 \tag{21}$$

This is the equation for the aeroelastic eigenvalues or natural frequencies of the structural-fluid system. In general these eigenvalues will be complex, $\omega = \omega_R + i\omega_I$. Formally by considering

$$w = \bar{w}e^{i\omega t} \tag{22}$$

or more rigorously by appealing to the calculus of residues and the Fourier Inversion Theorem we see that if any eigenvalue has a negative imaginary part

$$\omega_I < 0$$

the plate deflection will grow exponentially with time. Such motion is termed unstable. In general

$$\omega_R \neq 0$$

and the (dynamic) instability is termed 'flutter'. Occasionally $\omega_R \equiv 0$ and this (static) instability is termed 'divergence'. For $\omega_I = 0$ the motion is considered neutrally stable and for $\omega_I > 0$ the motion is stable. Unstable motion, of course, will lead to large and potentially dangerous motions (even in the nonlinear model) and these motions are of principal interest when they occur. The stability boundary is defined as the locus of curves in some parameter space for which $\omega_I \equiv 0$. Frequently this is expressed in terms of dynamic pressure, $q = \rho_\infty U_\infty^2/2$, *vs.* Mach number for a given structure.

Various types of aeroelastic instabilities

There are several types of flutter which may occur and these may be illustrated qualitatively by considering simplified approximations to Equation (16). Consider first only a single mode and equation, say $N = 1$. Then (21) becomes

$$M_1[\omega_1^2 - \omega^2] + \rho_\infty U_\infty^2 \bar{\bar{Q}}_{11}^M = 0 \tag{23}$$

Recall $\bar{\bar{Q}}_{11}^M$ is a complex number which varies with reduced frequency, $\omega a/U_\infty$ and Mach number (as well as panel geometry and natural mode shape). It turns out that $\bar{\bar{Q}}_{11}^M$ is an essentially negative real number, independent of reduced frequency, for subsonic flow, $M < 1$. On the other hand it is essentially an imaginary number which varies with frequency and may be either positive or negative for supersonic flow, $M > 1$. These results are for zero leading and trailing edge displacements which is the usual case in practice. However, see Appendix II for the effect of edge conditions.

Hence for subsonic flow ($\bar{\bar{Q}}_{11}^M = \bar{\bar{Q}}_{11_R}^M < 0$) we may compute the aeroelastic frequency as

$$\omega^2 = \omega_1^2 + \rho_\infty U_\infty^2 \frac{\bar{\bar{Q}}_{11}^M}{M_{11}} \tag{24}$$

Clearly if the right-hand side of (24) becomes less than zero then we will have an exponentially growing solution with no oscillation, i.e., the panel simply 'diverges'. The divergence condition is

$$(\rho_\infty U_\infty^2)_D = \frac{M_1 \omega_1^2}{-\bar{\bar{Q}}_{11_R}^M}$$

For supersonic flow the solution is more subtle. Separating real and imaginary parts of (23), we have ($\bar{Q}_{11}^M = i\bar{Q}_{11_I}^M$)

$$\begin{aligned} &\omega_1^2 - (\omega_R^2 - \omega_I^2) = 0 \\ &-M_1 2\omega_R\omega_I + \rho_\infty U_\infty^2 \bar{Q}_{11_I}^M = 0 \end{aligned} \tag{25}$$

For a neutrally stable solution, $\omega_I \equiv 0$ and hence

$$\begin{aligned} &\omega_R = \omega_1 \\ &\bar{Q}_{11}^M\left(\frac{\omega_R a}{U_{\infty_F}}, M\right) = \bar{Q}_{11}^M\left(\frac{\omega_1 a}{U_{\infty_F}}, M\right) = 0 \end{aligned} \tag{26}$$

Thus the instability is dynamic, 'flutter', with a frequency of oscillation equal to the modal natural frequency. Moreover, the velocity, U_{∞_F}, at which flutter occurs is that which permits the second of equations (26) to be satisfied. Of course, there may not be such a velocity for a given Mach number, e.g., in subsonic flow and also as it turns out at high supersonic flow, say $M > (2)^{\frac{1}{2}}$. Q_{11_I} may be thought of as a measure of the aerodynamic damping and hence this type of flutter is usually referred to as a 'single mode flutter due to negative damping'.

Flutter may also occur as a result of coupling between modes. This normally occurs at high supersonic Mach numbers (say $M > 2^{\frac{1}{2}}$) where $\bar{Q}_{11_I} \approx 0$ (and in any event $\bar{Q}_{11_I} > 0$ for all frequencies). Here $\bar{Q}_{12}^M = -\bar{Q}_{21}^M$ is essentially a purely real constant independent of frequency. For a two mode approximation, $N = 2$, (21) becomes

$$\begin{bmatrix} M_1(\omega_1^2 - \omega^2) & \rho_\infty U_\infty^2 \bar{Q}_{21_R}^M \\ -\rho_\infty U_\infty^2 \bar{Q}_{21_R}^M & M_2(\omega_2^2 - \omega^2) \end{bmatrix} = 0 \tag{27}$$

Solving for ω^2

$$\omega^2 = \frac{\omega_2^2 + \omega_1^2 \pm \left[(\omega_2^2 - \omega_1^2)^2 - 4\dfrac{\bar{Q}_{21}^2}{M_1 M_2}\rho_\infty^2 U_\infty^4\right]^{\frac{1}{2}}}{2} \tag{28}$$

For sufficiently large $\rho_\infty U_\infty^2$, the argument of the radical will become negative and at least one of the values of ω will have an imaginary part less than zero, i.e., the system flutters. Note that at the onset of flutter, when the radical vanishes, both eigenfrequencies have the same value. The two real structural natural frequencies, ω_1 and ω_2, which exist for $\rho_\infty U_\infty^2 = 0$ have merged due to aeroelastic coupling with increasing $\rho_\infty U_\infty^2$. The merging or coalescing of frequencies is typical of coupled mode 'flutter'.

In order to obtain quantitatively accurate results one must usually use more than one or two modes. Indeed there are certain structures for which many modes are needed, of the order ten or more. For example there are situations where the panel deformation takes the form of a traveling wave with exponential *spatial* growth. In such cases a large number of modes are required and indeed alternative methods of analysis may be preferable.

To illustrate typical results from flutter analyses we present a flutter boundary for a one-dimensional ($a/b = 0$), simply-supported plate in Fig. 2.4. (For the moment ignore the results for a square plate, $a/b = 1$.) Linearized, potential flow aerodynamic theory is used (see Chapter 4). A non-dimensional dynamic pressure $\lambda^* \equiv \rho_\infty U_\infty^2 a^3/D$ is plotted *vs.* Mach number for a fixed mass ratio, $\mu \equiv \rho_\infty a/\rho_m h$. For such a simple structure only these three non-dimensional aerodynamic parameters appear; for more complicated structures additional parameters will arise, see Chapter 6. As may be seen, the lowest dynamic pressure for the onset of 'flutter' occurs at $M = 1$. Indeed for this case the minimum flutter dynamic pressure is zero! More refined calculations including finite panel width or boundary layer aerodynamic effects will show that the flutter dynamic pressure is non-zero. In

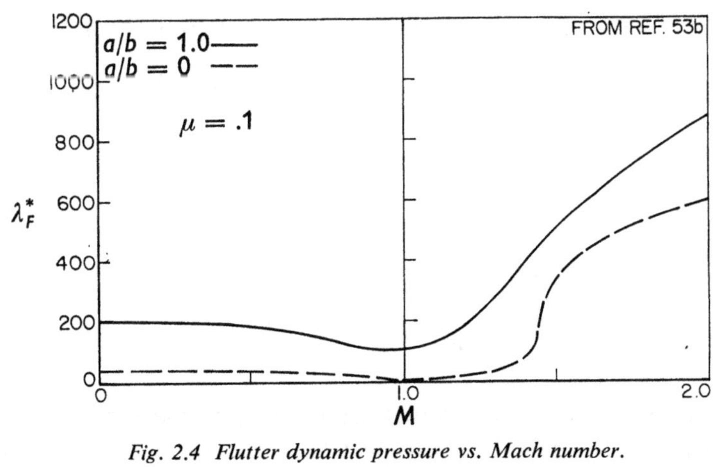

Fig. 2.4 Flutter dynamic pressure vs. Mach number.

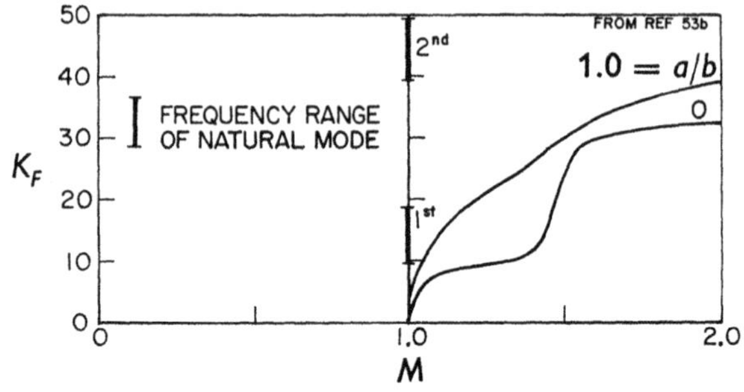

Fig. 2.5 Flutter frequency vs. Mach number.

Fig. 2.5 the corresponding flutter frequency, $K_f \equiv (ma^4/D)^{\frac{1}{2}}\omega_f$, is given. At high Mach number the flutter frequency is between the first and second panel natural modal frequencies; at moderate supersonic Mach number, $M = 1.1$–1.4, it is approximately equal to the first natural modal frequency; below $M = 1.1$ the flutter frequency rapidly falls to zero and the panel 'diverges' rather than flutters over the subsonic range of Mach number. These results confirm our earlier discussion of the various types of aeroelastic instabilities.

The experienced aeroelastician can, by an examination of the aerodynamic generalized forces, anticipate the type of 'flutter' or 'divergence' which may occur as indicated above. Alternatively the type of aeroelastic instability may be anticipated by considering simplified approximations to the aerodynamic pressure. For low Mach number subsonic flow (incompressible flow) the self-induced aerodynamic pressure is approximately given by

$$\Delta p^M \cong A_0 \rho_\infty U_\infty^2 \left[\frac{\partial^2 w}{\partial x^2} + \frac{2}{U_\infty} \frac{\partial^2 w}{\partial x \partial t} + \frac{1}{U_\infty^2} \frac{\partial^2 w}{\partial t^2} \right] \tag{29}$$

where

$$A_0 = \frac{1}{\pi} \int_{-x/a}^{1-x/a} \ln |\eta| \, d\eta$$

low Mach number, subsonic

The crucial term in the above is that proportional to panel curvature; it is the aerodynamic analog to the vertical load due to an applied in-plane structural load which can lead to 'buckling', or to use the aeroelastic term, 'divergence'. The second term is a form of gyroscopic damping and the third a form of fluid inertia loading. Neither are important for typical aeroelastic applications.

For supersonic flow at low reduced frequencies, $a\omega/U_\infty \ll 1$

$$\Delta p^M \cong \frac{\rho_\infty U_\infty^2}{(M^2-1)^{\frac{1}{2}}}\left[\frac{\partial w}{\partial x} + \frac{M^2-2}{M^2-1}\,\frac{1}{U_\infty}\,\frac{\partial w}{\partial t}\right] \tag{30}$$

low frequency, supersonic

At high Mach number the above reduces to the piston theory result. When $M \gg 1$ the first term is the important one, it leads to merging frequency flutter at sufficiently large dynamic pressure. At low supersonic Mach number the second term gives rise to negative damping when $M^2 < 2$ which leads to single mode flutter. At moderate (reduced) frequencies the above approximation breaks down and indeed a more exact aerodynamic analysis shows that the damping becomes positive again as the reduced frequency increases, i.e., an increase in physical frequency or decrease in airflow velocity. Nevertheless we can conclude from the above that for $M^2 > 2$, merging frequency flutter is likely, while for $M^2 < 2$ single mode flutter is a distinct possibility.

These approximations to the aerodynamic pressure and their derivation are discussed more fully in Chapter 4.

One can determine the consistency of this discussion with the earlier one using generalized forces by noting that

1. For $p \sim \partial^2 w/\partial x^2$

we have

$$\bar{Q}_{mn} \sim \int_0^a \frac{d^2\psi_m}{dx^2}\,\psi_n\,dx$$

When $\psi_m(0) = \psi_m(a) = 0$, then

$$\bar{Q}_{mn} \sim -\int_0^a \frac{d\psi_m}{dx}\,\frac{d\psi_n}{dx}\,dx$$

If we further assume the ψ_m are alternatively even and odd functions of x, then

$$\bar{\bar{Q}}_{22}, \bar{\bar{Q}}_{11} < 0 \quad \text{and} \quad \bar{\bar{Q}}_{12} = \bar{\bar{Q}}_{21} = 0$$

2. For $p \sim \partial w/\partial x$

we have

$$\bar{\bar{Q}}_{mn} \sim \int \frac{\mathrm{d}\psi_m}{\mathrm{d}x} \psi_n \mathrm{d}x$$

When $\psi_m(0) = \psi_m(a) = 0$, then

$$\bar{\bar{Q}}_{mn} = -\bar{\bar{Q}}_{nm}$$

thus

$$\bar{\bar{Q}}_{11} = \bar{\bar{Q}}_{22} = 0$$

$$\bar{\bar{Q}}_{21} = -\bar{\bar{Q}}_{12}$$

3. For $p \sim \partial w/\partial t$

we have

$$\bar{\bar{Q}}_{mn} \sim \int \psi_m \psi_n \mathrm{d}x$$

For orthogonal modes

$$\bar{\bar{Q}}_{mn} = 0 \text{ if } m \neq n$$
$$> 0 \text{ if } m = n$$

It is important to note that the 'piston theory' aerodynamic approximation (which is the most popular in the literature) only predicts the high Mach number results accurately. Indeed it gives no information at all about subsonic divergence or single-degree-of-freedom flutter.

For further study of linear models the reader may consult several basic sources. Virtually all have used the eigenvalue frequency method with various methods for treating the spatial domain. Hedgepeth [28] wrote a seminal paper on the modal approach using piston theory aerodynamics. This approach was subsequently developed using the full linearized potential flow aerodynamic model as is thoroughly discussed by Cunningham [29]. There are alternatives to the modal approach, of course, although none has been as widely used to date. There is, for example, the finite element method as discussed by Olson [30]. Finite difference methods have also been suggested but infrequently used. Among the more interesting methods for treating the spatial domain are so-called 'exact' solution techniques which involve no further approximation in terms of finite numbers of modes or elements. Using the piston theory aerodynamics, Hedgepeth, Houbolt and Movchan obtained results using the classical separation of variable method. This approach is lucidly presented by Dugundji [31]. Goland and Luke, Zeydel and, most recent and completely, Yates [32] have used a Laplace Transform (with respect to the streamwise coordinate) approach while employing the full linearized potential flow theory which has given interesting results. Spriggs, Messiter and Anderson [33] have developed an inner-outer expansion technique which is advantageous for considering certain difficult limiting cases involving large tension loads or long narrow panels.

In the early work on this problem there was considerable discussion of the 'membrane' paradox and whether or not the flutter motion took the form of a 'standing wave' or 'traveling wave'. The former was associated with the fact that a membrane (plate under tension with no bending stiffness) does not flutter according to an exact analysis using piston theory aerodynamics. Hence no merging frequency flutter at high Mach number is possible. Single mode flutter at transonic – low supersonic speeds and subsonic divergence are possible, of course. The 'paradox' arises from the fact for a plate of finite bending stiffness under a tension load, a reduction in bending stiffness leads to a decrease in flutter dynamic pressure. Hence one would expect a finite flutter dynamic pressure as the bending stiffness decreases to zero. In fact this is true as Spriggs, *et al.* [33], have shown. The problem from a mathematical point of view is of singular perturbation type and requires a careful distinction between the limit of bending/tension stiffness approaching zero due to a reduction in bending stiffness on the one hand or an increase in tension on the other. In the latter case the conventional analysis applies and the flutter dynamic pressure increases with increasing tension and approaches infinity as the tension becomes indefinitely large. In the former case there is in fact a finite flutter dynamic pressure as the bending stiffness vanishes. The panel deflection shape at flutter, 'flutter mode', is an exponentially growing wave with a peak deflection in a structural 'boundary layer' near the panel trailing edge. It is this characteristic which makes the problem one requiring an inner-outer expansion analysis appropriate to singular perturbation problems of this type.

As discussed in [34] there is a mathematical analogy between the membrane problem and a long narrow plate of finite bending stiffness (with or without tension load). Thus the membrane analysis may be used for such panels and, in particular, the structural 'boundary layer' character of the flutter mode is present. It is this boundary layer characteristic which explains the failure of classical, constant amplitude 'traveling wave' analyses for long, narrow plates even though the characteristic wavelength of the flutter is on the order of the plate width and hence small compared to the panel length for such plates. For low supersonic and subsonic Mach numbers, the flutter mode is not of boundary layer type and a 'traveling wave' analysis gives results in reasonable agreement with more elaborate analyses which account for the effects of finite panel length. See [34] and also the discussion of Yates [32]. Also consult [35] and [36] for a detailed discussion of 'traveling wave' analyses.

2.2 General models

In the linear approximation one may formulate a rather general theoretical model which is valid for arbitrary structural shapes, support conditions, and stiffness distributions (orthotropicity) [37]. Perhaps the most economical and informative way to approach the problem is to employ an energy approach using Hamilton's Principle. Considering first, for definiteness, an orthotropic flat plate we may write the elastic potential energy (using rectangular cartesian coordinates)

$$U = \tfrac{1}{2}\iint\left[D_x\left(\frac{\partial^2 w}{\partial x^2}\right)^2 + D_y\left(\frac{\partial^2 w}{\partial y^2}\right)^2 + 2D_1\,\frac{\partial^2 w}{\partial x^2}\,\frac{\partial^2 w}{\partial y^2} + 4D_{xy}\left(\frac{\partial^2 w}{\partial x \partial y}\right)^2\right]\mathrm{d}x\,\mathrm{d}y \tag{1}$$

where for an isotropic plate $D_x = D_y = D$, $D_1 = \nu D$ and $D_{xy} = [(1 - \nu)D]/2$, the kinetic energy

$$T = \tfrac{1}{2}\iint m\left(\frac{\partial w}{\partial t}\right)^2 \mathrm{d}x\,\mathrm{d}y \tag{2}$$

and the virtual work done by the pressure distribution

$$\delta W = -\iint \Delta p\,\delta w\,\mathrm{d}x\,\mathrm{d}y \tag{3}$$

Hamilton's Principle may be stated as

$$\delta\int (T - U)\mathrm{d}t + \int \delta W \mathrm{d}t = 0 \tag{4}$$

Applying the variational or virtual operator δ we may derive the Euler-Lagrange equation of equilibrium *and* associated boundary conditions. For a rectangular, isotropic plate with the classical support conditions we obtain as the Euler-Lagrange equation

$$D\nabla^4 w + m\frac{\partial^2 w}{\partial t^2} + \Delta p = 0 \tag{5}$$

with associated boundary conditions

on $x = \text{constant}$

$$w = 0 \quad \text{or} \quad \frac{\partial^3 w}{\partial x^3} + (2 - \nu)\frac{\partial^3 w}{\partial x \partial y^2} = 0$$

$$\frac{\partial w}{\partial x} = 0 \quad \text{or} \quad \frac{\partial^2 w}{\partial x^2} + \nu\frac{\partial^2 w}{\partial y^2} = 0$$

on $y = \text{constant}$

$$w = 0 \quad \text{or} \quad \frac{\partial^3 w}{\partial y^3} + (2 - \nu)\frac{\partial^3 w}{\partial y \partial x^2} = 0$$

$$\frac{\partial w}{\partial y} = 0 \quad \text{or} \quad \frac{\partial^2 w}{\partial y^2} + \nu\frac{\partial^2 w}{\partial x^2} = 0$$

on $x =$ constant, $y =$ constant (a corner)

$$w = 0 \quad \text{or} \quad \frac{\partial^2 w}{\partial x \partial y} = 0 \tag{6}$$

One can derive similar relations for non-rectangular, orthotropic plates with more complicated support conditions. However, it is generally preferable to proceed by using a Rayleigh-Ritz modal approximation starting from the energy and work relations directly.

Expanding the plate deflection as

$$w = \sum_m q_m(t)\psi_m(x, y) \tag{7}$$

and substituting into (1), (2) and (3) we have

$$U = \tfrac{1}{2} \sum_{mn}\sum q_m q_n K_{mn} \tag{8}$$

$$T = \tfrac{1}{2} \sum_{mn}\sum \dot{q}_m \dot{q}_n M_{mn} \tag{9}$$

$$\delta w = - \sum_n Q_n \delta q_n \tag{10}$$

where

$$K_{mn} = \iint \left[D_x \frac{\partial^2 \psi_m}{\partial x^2} \frac{\partial^2 \psi_n}{\partial x^2} + D_y \frac{\partial^2 \psi_m}{\partial y^2} \frac{\partial^2 \psi_n}{\partial y^2} + 2D_1 \frac{\partial^2 \psi_m}{\partial x^2} \frac{\partial^2 \psi_n}{\partial y^2} + 4D_{xy} \frac{\partial^2 \psi_m}{\partial x \partial y} \frac{\partial^2 \psi_n}{\partial x \partial y} \right] \mathrm{d}x\,\mathrm{d}y \tag{11}$$

$$M_{mn} = \iint m \psi_m \psi_n \,\mathrm{d}x\,\mathrm{d}y \tag{12}$$

$$Q_n = \iint \Delta p \psi_n \,\mathrm{d}x\,\mathrm{d}y \tag{13}$$

Defining a Lagrangian, $L \equiv T - U$, the Euler-Lagrange or simply Lagrange's equations are:

$$\frac{\mathrm{d}}{\mathrm{d}t}\left(\frac{\partial L}{\partial \dot{q}_n} \right) - \frac{\partial L}{\partial q_n} + Q_n = 0$$

or

$$\sum_m M_{mn}\ddot{q}_m + K_{mn}q_m + Q_n = 0 \tag{14}$$

The essential advantage of the Rayleigh-Ritz approach over Galerkin's method is the choice of functions ψ_m. In Galerkin's method all boundary conditions must be satisfied by each ψ_m; in the Rayleigh-Ritz approach only the geometric boundary conditions need be satisfied by each ψ_m. In fact as we shall discuss below, one can modify the method so that the ψ_m need not satisfy any of the boundary or support conditions.

Equations (14) can be simplified significantly if we are so fortunate as to know and use the panel natural modes. Then

$$M_{mn} = M_n \delta_{mn}$$
$$K_{mn} = \omega_n^2 M_n \delta_{mn}$$

and we have

$$M_n[\ddot{q}_n + \omega_n^2 q_n] + Q_n = 0 \tag{15}$$

Of course, by considering (14) for $Q_n \equiv 0$ we can determine the eigenfunctions as a linear combination of any set of admissible primitive functions, ψ_m, by solving the eigenvalue problem

$$[-\omega^2[M_{mn}] + [K_{mn}]]\{q_n\} = \{0\} \tag{16}$$

Other techniques could be used for determining the eigenmodes such as the finite element or finite difference methods. For odd shaped panels, in particular, such methods may be advantageous for first determining the natural modes before formulating the aeroelastic equations of motion in terms of such modes.

Arbitrary support conditions

Let us consider a modification to the Rayleigh-Ritz method which allows one to use a single modal expansion for any support condition, thereby eliminating one of the principal practical disadvantages of the modal expansion approach. We shall distinguish between elastic supports and rigid supports; surprisingly perhaps, the former is conceptually easier to handle than the latter. For example, consider a point translational spring support with stiffness, K_T, at some point, x_s, y_s. The associated elastic energy is

$$\tfrac{1}{2}K_T w^2(x_s, y_s) \tag{17}$$

which leads to an added term to the generalized stiffness of the form

$$K_{mn} = \ldots + K_T \psi_m(x_s, y_s)\psi_n(x_s, y_s) \tag{18}$$

Other types of stiffness or mass supports may similarly be taken into account. Using several point supports, a line support condition may be approximated or the line support may be treated directly by analogous means.

By allowing $K_T \to \infty$, one may obtain a rigid support, however there may be a practical difficulty with taking the limit as $K_T \to \infty$ or similar limits as the corresponding stiffness terms create numerical problems due to small differences between large numbers. This can be overcome by careful numerical work but an alternative, more efficient, and possibly more elegant approach is available using Lagrange multipliers and appropriate constraint relations. (This approach may also be used for elastic supports but there is no particular net advantage.) For example, consider a rigid translational support at points x_s, y_s. The constraint conditions are

$$w(x_s, y_s) = \sum_m q_m \psi_m(x_s, y_s) = 0 \qquad s = 1, \ldots, S \tag{19}$$

The Lagrangian, $L \equiv T - U$, is modified to read

$$L \equiv T - U + \sum_s \lambda_s \sum_m q_m \psi_m(x_s, y_s) \tag{20}$$

and Lagrange's equations become

$$\frac{\mathrm{d}}{\mathrm{d}t}\left(\frac{\partial L}{\partial \dot{q}_n}\right) - \frac{\partial L}{\partial q_n} + Q_n = 0$$

or

$$\sum_{m=1}^{N} M_{nm}\ddot{q}_m + \tilde{K}_{mn}q_m + \sum_s \lambda_s \psi_n(x_s, y_s) + Q_n = 0 \qquad n = 1, \ldots, N \tag{21}$$

$$\tilde{K}_{mn} \equiv (K_{mn} + K_{nm})/2$$

and

$$\frac{\partial L}{\partial \lambda_s} = 0 \qquad s = 1, \ldots, S \tag{22}$$

(22) simply reproduces (19), of course. Equations (21) and (22) are $N + S$ equations for the N, q_m, plus S, λ_s, unknowns. In solving these equations, by step-by-step temporal integration for example, one must use the fact that Equations (19) constrain the modal velocities, $\dot{q}_m$, and accelerations, $\ddot{q}_m$, as follows

$$\dot{w}(x_s, y_s) = \sum_m \dot{q}_m \psi_m(x_s, y_s) = 0 \tag{23}$$

$$\ddot{w}(x_s, y_s) = \sum_m \ddot{q}_m \psi_m(x_s, y_s) = 0 \qquad s = 1, \ldots, S \tag{24}$$

Hence, one can only specify initial conditions on $N - S$ of the $\dot{q}_m$ and q_m, solving for the remaining S from Equations (19) (or (22)) and (23). One may then solve for all of the N, $\ddot{q}_m$, and S, λ_s, from Equations (21) and (24). With this information one may project to the next time step and obtain a time history of the plate motion.

In Fig. 2.4 and 2.5 results are presented for flutter dynamic pressure and frequency *vs.* Mach number for one-dimensional ($a/b = 0$) and square ($a/b = 1$) simply-supported plates. Note the results are qualitatively similar for the two panel length to width ratios except near $M = 1$.

To illustrate the importance of support conditions we examine the flutter of a square panel on four symmetrically placed rigid point supports. For simplicity we consider high Mach numbers where piston theory aerodynamics are appropriate and hence the three aerodynamic parameters λ^*, μ, M may be reduced to two, λ^*/M and μ/M. In Fig. 2.6 flutter dynamic pressure is plotted *vs.* position of the supports along a panel diagonal. As the four supports are moved from the panel corners to the center, λ_F^*/M first increases then decreases. Moreover the flutter frequency becomes zero as the supports approach the center leading to divergence rather than flutter even at high supersonic Mach number. Clearly, support conditions can be qualitatively as well as quantitatively important.

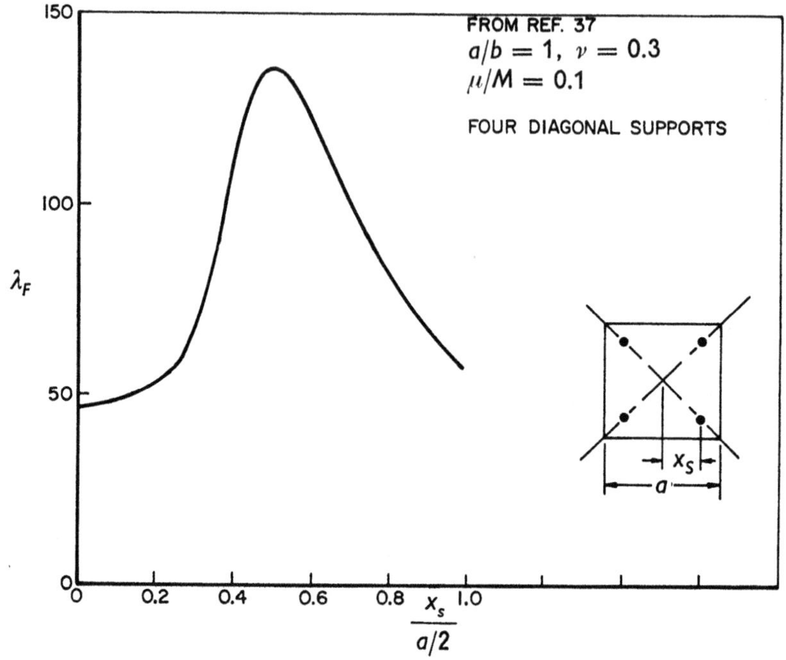

Fig. 2.6 Flutter dynamic pressure vs. support position.

Effects of structural damping

Although some theoretical estimates of damping are available for plates constructed from viscoelastic materials, the source of damping in a nominally elastic plate is more obscure. Hence the practice has usually been to add to the Euler-Lagrange equations (14) empirical terms of the form

$$\beta_n \delta_{mn} \dot{q}_n$$

If plate normal modes have been used (or frequently even if they have not!) β_n is expressed as

$$\beta_n = 2\zeta_n \omega_n$$

where ω_n is the n-th natural mode frequency and ζ_n is a non-dimensional damping coefficient which is usually taken to lie between .005–.05. Structural damping is not normally important for flutter of plates and shells except of the single mode type may be but important for resonant response to external fluctuating pressures. For single mode flutter the effect of fluid boundary layer contributions to aerodynamic damping may also be important. See Chapter 4.

Other structural parameters

There are several other structural parameters which may be of importance, among these are pressurization, in-plane loads (mechanical or thermal) and curvature including geometrical imperfections. A static pressure loading has no effect (on a flat plate) within the framework of a linear theoretical model and in-plane loads are of greatest significance when they lead to (nonlinear) buckling. Hence these effects will be discussed in Chapter 3.

The effect of curvature is of significance even in the linear model and may be formally included as a modification to the generalized stiffness K_{mn}. Because of the mathematical similarity in the curvature and nonlinear bending effects we delay an explicit evaluation of the curvature effect until our discussion of nonlinear models. With the modified K_{mn} the effect of curvature may be included as per the above discussion.

For further study the reader may consult several sources. Erickson has discussed orthotropic [38] and sandwich [39] panels, Gaspers and Redd [40] the effects of orthotropicity and angle of flow, Bohon and Anderson [41] orthotropicity and the importance of boundary support conditions and Shore [42] the effect of structural damping. Voss [43] considered cylindrical shell flutter using a rather complete set of shell equations attributed to Goldenweiser. Later authors have tended to use the somewhat simpler Donnell type shell equations; see for example Olson and Fung [44] and the work of Dixon and Hudson [45] (who treated conical shells). All of the above authors have used piston theory aerodynamics. LeClerc [46] and Johns [47] have recently used the full potential flow theory for flutter analyses of cylindrical shells.

3

Nonlinear theoretical aeroelastic models

3.1 Illustrative model

The principal source of structural nonlinearity is the tension which results from in-plane stretching of the panel. This stretching or change in length of the panel is a consequence of large out-of-plane bending. In the simplest approximation the one-dimensional plate equation is

$$D\frac{\partial^4 w}{\partial x^4} - N_x\frac{\partial^2 w}{\partial x^2} + m\frac{\partial^2 w}{\partial t^2} + \Delta p = 0 \tag{1}$$

where N_x is the tension induced by stretching. It may be evaluated by considering the strain or relative change in length, $\Delta a/a$, of the panel. From elementary calculus of the deformed panel geometry,

$$\frac{\Delta a}{a} = \frac{\int_0^a \left[1 + \left(\frac{\partial w}{\partial x}\right)^2\right]^{\frac{1}{2}} \mathrm{d}x - a}{a}$$

$$\cong \frac{1}{2}\frac{\int_0^a \left(\frac{\partial w}{\partial x}\right)^2 \mathrm{d}x}{a}$$

where the last line follows from a binomial expansion valid for

$$\left(\frac{\partial w}{\partial x}\right)^2 \ll 1$$

As we shall see the slope of the panel for a flat plate is on the order of the thickness/length and for a curved plate on the order of rise height/length. For the former case the slope is always very small and for the latter it is still usually substantially less than one. The tension is evaluated as the stress times the cross-sectional area (for unit width). Using Hooke's Law for an elastic material and the previous expression for strain

$$N_x = E\frac{\Delta a}{a}h = \frac{Eh}{2a}\int_0^a \left(\frac{\partial w}{\partial x}\right)^2 \mathrm{d}x \tag{2}$$

Substituting (2) into (1) gives

$$D\frac{\partial^4 w}{\partial x^4} - \left[\frac{Eh}{2a}\int_0^a \left(\frac{\partial w}{\partial x}\right)^2 \mathrm{d}x\right]\frac{\partial^2 w}{\partial x^2} + m\frac{\partial^2 w}{\partial t^2} + \Delta p = 0 \tag{3}$$

Hence the tension increases as the deflection increases (note the tension is always positive for any deflection). It is this tension which limits the plate to finite deflections even when the motion is unstable in the linear model. For large external pressure fluctuations it may also significantly reduce the plate response over that predicted by the linear theory even when the motion is stable.

The question arises how large must the deflection be before the nonlinear tension becomes significant? One may reason that this will occur when the (linear) bending and (nonlinear) tension stiffness terms are of comparable magnitude. Assuming that the scale of the spatial derivatives is

$$\frac{\partial}{\partial x} \sim 1/a$$

then the bending term is of the order

$$Eh^3\frac{w}{a^4}$$

and the tension term of order

$$Eh\frac{w^3}{a^4}$$

If these terms are of the same magnitude then

$$w \sim h$$

For $w \ll h$, the bending term is more important than the tension term and conversely for $w \gg h$. One may further argue that near the onset of flutter the inertial, aerodynamic and stiffness terms must all be of comparable magnitude and hence that near but beyond the stability boundary the plate limit cycle flutter amplitude will be of the order of the plate thickness, h. Similar but more sophisticated reasoning will lead to the conclusion that for a curved plate, the flutter amplitude will be on the order of the maximum rise height just beyond the stability boundary [20].

To assess the effects of structural nonlinearity on a stable panel responding to external pressures we compare the size of the loading

$$\Delta p^E$$

to the magnitude of the stiffness terms. Ignoring resonance effects we ask when does

$$\Delta p^E \overset{?}{\sim} Eh^3 \frac{w}{a^4}$$

or

$$\frac{\Delta p^E a^4}{Eh^4} \overset{?}{\sim} w/h$$

Clearly if

$$\frac{\Delta p^E a^4}{Eh^4}$$

is of the order of unity or greater structural nonlinearities will be significant. Moreover, if there is substantial frequency matching between the flow and structure (resonance effects) then this parameter may be such smaller and nonlinearities may still be important. Note this parameter is also the one of significance for static external pressure loadings. Static loadings of course do not lead directly to dynamic response but indirectly they may have a significant effect on dynamic response if they are sufficiently large to deform the structure into its nonlinear stiffness regime.

With these simple ideas in mind we are now in a position to begin a systematic discussion of nonlinear structural theories of plates and shells. The mathematics will become more elaborate but the basic physical concepts will be essentially the same. We could of course, develop nonlinear modal equations of motion from (3) by expansion of w, *viz.*

$$w = \sum_m q_m(t)\psi_m(x, y)$$

and application of Galerkin's method [20]. For the sake of brevity we omit this calculation and proceed directly to more general models.

3.2 General models

We begin with the basic flat plate model and then consider in turn various additions to it.

Nonlinear structural model for a flat plate

The equation of motion for a two-dimensional, isotropic flat plate may be written in a form attributed to von Karman

$$D\nabla^4 w = \frac{\partial^2 F}{\partial y^2}\frac{\partial^2 w}{\partial x^2} + \frac{\partial^2 F}{\partial x^2}\cdot\frac{\partial^2 w}{\partial y^2} - 2\frac{\partial^2 F}{\partial x \partial y}\frac{\partial^2 w}{\partial x \partial y} - m\frac{\partial^2 w}{\partial t^2} - \Delta p \quad (1)$$

$$\frac{\nabla^4 F}{Eh} = \left(\frac{\partial^2 w}{\partial x \partial y}\right)^2 - \left(\frac{\partial^2 w}{\partial x^2}\right)\left(\frac{\partial^2 w}{\partial y^2}\right) \quad (2)$$

Here F is the Airy stress potential function. The in-plane stress resultants (tensions and shears) are obtained from

$$N_x = \frac{\partial^2 F}{\partial y^2}$$

$$N_y = \frac{\partial^2 F}{\partial x^2} \quad (3)$$

$$N_{xy} = -\frac{\partial^2 F}{\partial x \partial y} \tag{3}$$

Note that the introduction of F allows the in-plane equations of equilibrium to be identically satisfied

$$\begin{aligned} \frac{\partial N_x}{\partial x} + \frac{\partial N_{xy}}{\partial y} &= 0 \\ \frac{\partial N_y}{\partial y} + \frac{\partial N_{xy}}{\partial x} &= 0 \end{aligned} \tag{4}$$

In the above the in-plane inertias have been neglected as is usual on the basis of the high natural frequencies (stiffness) of the in-plane motion as compared to the out-of-plane (bending) motion.

Equation (2) is essentially a strain compatibility condition which follows from the relations between three in-plane stress resultants, N_x, N_y, N_{xy}, and two in-plane displacements, u, v,

$$N_x = \frac{Eh}{(1-\nu^2)}[\varepsilon_{xo} + \nu\varepsilon_{yo}] \qquad N_{xy} = Gh\varepsilon_{xyo}$$

$$N_y = \frac{Eh}{(1-\nu^2)}[\varepsilon_{yo} + \nu\varepsilon_{xo}]$$

where

$$\begin{aligned} \varepsilon_{xo} &= \frac{\partial u}{\partial x} + \frac{1}{2}\left(\frac{\partial w}{\partial x}\right)^2 \\ \varepsilon_{yo} &= \frac{\partial v}{\partial y} + \frac{1}{2}\left(\frac{\partial w}{\partial y}\right)^2 \\ \varepsilon_{xyo} &= \frac{\partial v}{\partial x} + \frac{\partial u}{\partial y} + \frac{\partial w}{\partial x}\frac{\partial w}{\partial y} \end{aligned} \tag{5}$$

A derivation of Equations (1–5) via Hamilton's Principle is given in Appendix I for the sake of completeness.

In addition to the equations of equilibrium we must also consider boundary or support conditions. Consider a rectangular plate of dimensions a and b. Recall Fig. 2.1.

On a particular edge, say $x = 0$, the classical boundary conditions usually considered are

$$
\begin{aligned}
w &= 0 \quad \text{or} \quad V = 0 \\
\frac{\partial w}{\partial x} &= 0 \quad \text{or} \quad M = 0 \\
u &= 0 \quad \text{or} \quad N_x = 0 \\
v &= 0 \quad \text{or} \quad N_y = 0
\end{aligned}
\tag{6}
$$

where V and M are the effective shear and bending moment respectively which can be expressed in terms of w. If the boundary conditions on u, v are appropriate for a given application it may be preferable to deal with in-plane equations of equilibrium obtained by substituting Equations (5) into (4) to obtain two equations for u and v in terms of w.

For other more complicated boundary conditions involving flexible supports or interior supports, more complicated plate geometries or stiffness distributions, one may wish to deal directly with energies via Hamilton's Principle. We shall turn to these matters subsequently.

Now let us consider a solution to Equations (1–6) using again as in Chapter 2 a modal expansion technique. For simplicity we shall consider the in-plane boundary conditions to be

$$
\begin{aligned}
N_x &= N_{xy} = 0 \quad \text{on} \quad x = 0, a \\
N_y &= N_{xy} = 0 \quad \text{on} \quad y = 0, b
\end{aligned}
\tag{7}
$$

and the out-of-plane boundary conditions to be

$$
w = 0 \tag{8}
$$

and either

$$
\frac{\partial w}{\partial x} = 0 \quad \text{or} \quad M_x = 0 \quad \text{on} \quad x = 0, a
$$

$$\frac{\partial w}{\partial y} = 0 \quad \text{or} \quad M_y = 0 \quad \text{on} \quad y = 0, b$$

Expand w and F

$$w = \sum_m q_m \psi_n(x, y)$$
$$F = \sum_r a_r \phi_r(x, y) \tag{9}$$

where each ϕ_r, ψ_m satisfy (7) and (8) respectively and substitute into (1) and (2). Applying the Galerkin averaging procedure we obtain

$$\sum_m q_m K_{mn} + \ddot{q}_m M_{mn} + Qn = \sum_r \sum_m a_r q_m L_{rmn} \tag{10a}$$

$$\sum_r a_r A_{rs} = \sum_l \sum_m q_l q_m B_{lms} \tag{10b}$$

where

$$M_{mn} = \iint m\psi_m \psi_n \,\mathrm{d}x\,\mathrm{d}y$$
$$K_{mn} \equiv \iint D\nabla^4 \psi_m \psi_n \,\mathrm{d}x\,\mathrm{d}y$$
$$Q_n \equiv \iint \Delta p \psi_n \,\mathrm{d}x\,\mathrm{d}y$$
$$A_{rs} \equiv \iint \frac{\nabla^4 \phi_r}{Eh} \phi_s \,\mathrm{d}x\,\mathrm{d}y \tag{11}$$
$$B_{lms} \equiv \iint \left[\frac{\partial^2 \psi_l}{\partial x \partial y} \frac{\partial^2 \psi_m}{\partial x \partial y} - \frac{\partial^2 \psi_l}{\partial x^2} \frac{\partial^2 \psi_m}{\partial y^2} \right] \phi_s \,\mathrm{d}x\,\mathrm{d}y$$
$$L_{rmn} \equiv \iint \left[\frac{\partial^2 \phi_r}{\partial y^2} \frac{\partial^2 \psi_m}{\partial x^2} + \frac{\partial^2 \phi_r}{\partial x^2} \frac{\partial^2 \psi_m}{\partial y^2} - 2 \frac{\partial^2 \phi_r}{\partial x \partial y} \frac{\partial^2 \psi_m}{\partial x \partial y} \right] \psi_n \,\mathrm{d}x\,\mathrm{d}y$$

Equations (10) are a system of nonlinear coupled, ordinary differential equations in time which may be solved by several approximate techniques.

Note that the second of these is only an algebraic equation. As we have seen before the aerodynamic motion forces will involve memory integrals of the plate modal amplitudes, q_m; however we defer a detailed discussion of the aerodynamic forces until Chapter 4.

To illustrate the results obtained from such analyses the limit cycle response (peak panel deflection normalized by panel thickness) of a fluttering, simply-supported plate is plotted in Fig. 3.1 and 3.2 *vs.* dynamic pressure.

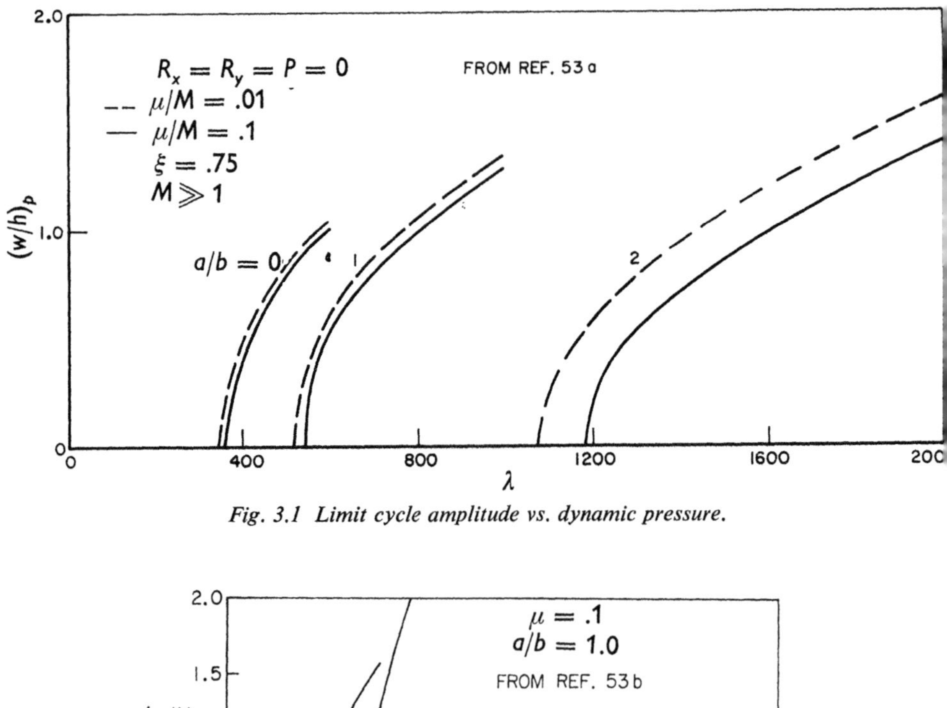

Fig. 3.1 Limit cycle amplitude vs. dynamic pressure.

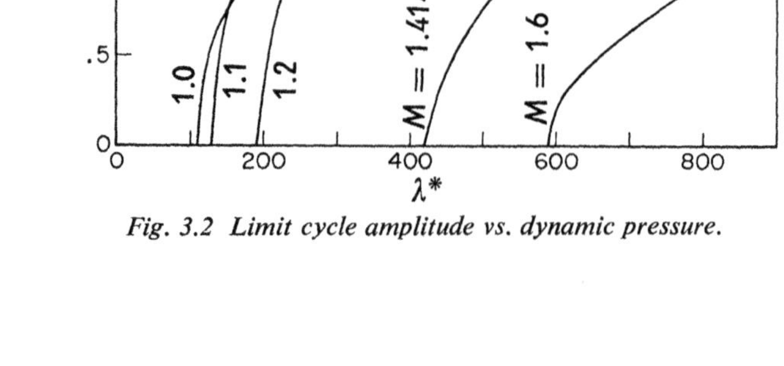

Fig. 3.2 Limit cycle amplitude vs. dynamic pressure.

Results are shown for several length to width ratios at high Mach number and a square plate at various Mach number. The in-plane displacements are taken as zero on the panel edges. The results for various length to width ratios are qualitatively similar but there appears to be a tendency for larger panel response at the lower Mach numbers.

We shall next turn to various structural complications beyond the above flat plate model.

Effects of static external loadings

There are two principal sources of such loadings, a static pressure differential, Δp^{ES}, and static in-plane loads, either mechanical or thermal. The former is readily incorporated into our analysis as previously discussed leading to a term of the form

$$Q_n^{ES} = \iint \Delta p^{ES} \psi_n \mathrm{d}x \mathrm{d}y \tag{12}$$

The latter is also not difficult to include as our problem is linear in the in-plane deformations and stresses *per se*. Hence any additional in-plane loads due to mechanical or thermal forces may be superposed with those due to (nonlinear) bending. Thus if we have specified static external in-plane loads, N_x^{ES}, N_y^{ES}, N_{xy}^{ES}, we add the following terms to the right-hand side of Equation (1).

$$N_x^{ES} \frac{\partial^2 w}{\partial x^2} + N_y^{ES} \frac{\partial^2 w}{\partial y^2} + 2N_{xy}^{ES} \frac{\partial^2 w}{\partial x \partial y} \tag{13}$$

For a constant temperature differential between plate and in-plane boundary supports

$$N_x^{ES} = N_y^{ES} = -\frac{Eh\alpha \Delta T}{(1-\nu)} \tag{14}$$

where ΔT is the temperature differential and α the coefficient of thermal expansion. (14) are the well-known relations between thermal stress and temperature differential. For a variable temperature differential it may be more convenient to supplement the right-hand side of Equation (2) by the term

$$-\alpha\nabla^2(\Delta T)$$

The effect of ΔT is now implicit in Equation (1) through F and no explicit use of N_x^{ES}, etc., is necessary.

All of the above steps follow rigorously from an inclusion of

$$\begin{aligned} &N_x^{ES} \quad \text{or} \quad -\frac{Eh\alpha\Delta T}{(1-\nu)} \\ &N_y^{ES} \quad \text{or} \quad -\frac{Eh\alpha\Delta T}{(1-\nu)} \\ &N_{xy}^{ES} \quad \text{or} \quad 0 \end{aligned} \tag{15}$$

on the right sides of Equations (5) and a rederivation of Equations (1) and (2) following the same procedure as before. See Appendix I.

Finally the modal stiffness coefficients, Equations (11), are modified by adding terms

$$K_{mn} = \ldots + \iint \left[N_x^{ES} \frac{\partial^2 \psi_m}{\partial x^2} + N_y^{ES} \frac{\partial^2 \psi_m}{\partial y^2} + 2N_{xy}^{ES} \frac{\partial^2 \psi_m}{\partial x \partial y} \right] \psi_n \, \mathrm{d}x \, \mathrm{d}y$$

To illustrate the effect of static external pressure loading consider Fig. 3.3 and 3.4. The former gives the peak of the oscillating dynamic panel response which occurs about a static deformed shape as the panel flutters under a given static pressure loading. As the loading is increased the dynamic panel response decreases (the static response, which is not shown, continues to increase). At the point of zero dynamic response the panel ceases to flutter and a flutter boundary may be constructed from such results as shown in Fig. 3.4 in terms of dynamic pressure *vs.* static pressure loading.

The effect of in-plane static loadings is even more complex. Again detailed panel response may be determined and from such results one may map out various panel response regions in terms of dynamic pressure *vs.* in-plane load. In Fig. 3.5 we see response regions of several different types including flat, stable panel; buckled panel; fluttering panel with either simple harmonic or more complicated periodic limit cycle motion.

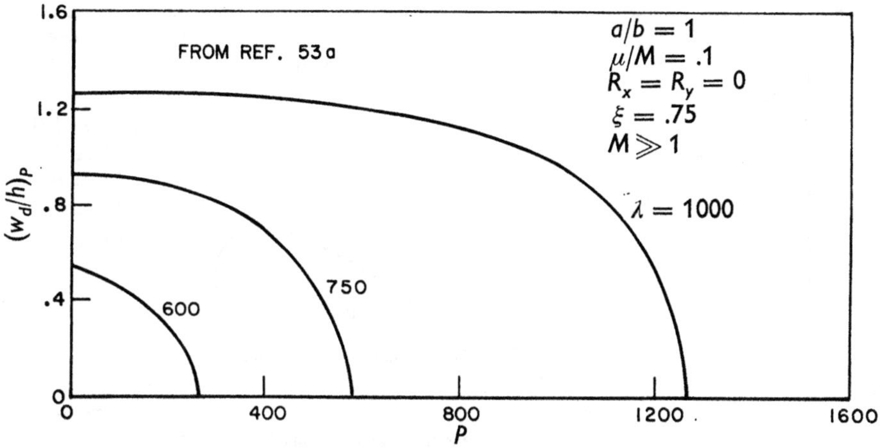

Fig. 3.3 Dynamic limit cycle amplitude vs. static pressure differential.

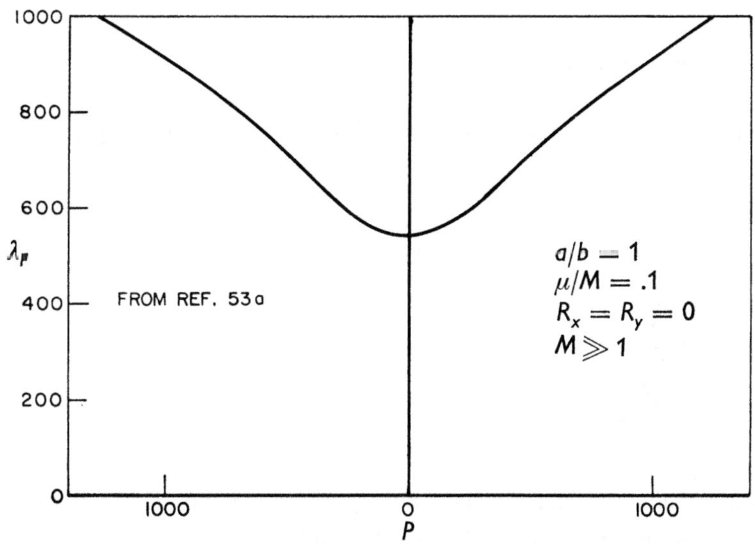

Fig. 3.4 Flutter dynamic pressure vs. static pressure differential.

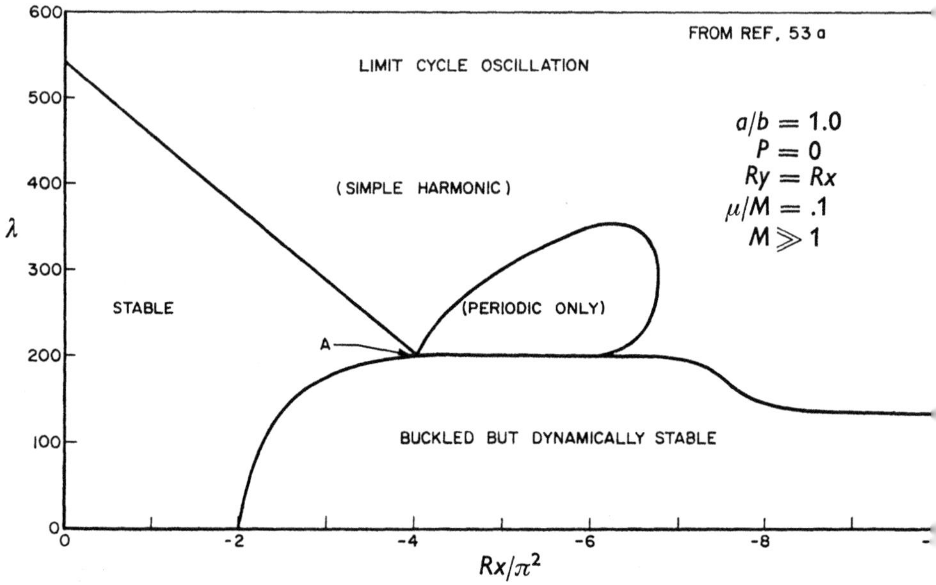

Fig. 3.5 Stability regions for plate with in-plane load.

Effects of curvature

Consider a curved plate whose initial undeformed shape is given by

$$z_p = z_p(x, y) \tag{16}$$

If the plate is one of shallow curvature, i.e.,

$$\left(\frac{\partial z_p}{\partial x}\right)^2 \ll 1$$

$$\left(\frac{\partial z_p}{\partial y}\right)^2 \ll 1$$

then we may consider the right-hand sides of (1) and (2) to still apply where we replace w by $w + z_p$. Hence we add to (1)

$$\frac{\partial^2 F}{\partial y^2}\,\frac{\partial^2 z_p}{\partial x^2} + \frac{\partial^2 F}{\partial x^2}\,\frac{\partial^2 z_p}{\partial y^2} - 2\,\frac{\partial^2 F}{\partial x \partial y}\,\frac{\partial^2 z_p}{\partial x \partial y}$$

and to (2)

$$2\frac{\partial^2 z_p}{\partial x \partial y}\,\frac{\partial^2 w}{\partial x \partial y} - \frac{\partial^2 z_p}{\partial x^2}\,\frac{\partial^2 w}{\partial y^2} - \frac{\partial^2 z_p}{\partial y^2}\,\frac{\partial^2 w}{\partial x^2}$$

Note that in the latter there are no terms like

$$\left(\frac{\partial^2 z_p}{\partial x \partial y}\right)^2$$

since strain is measured from the undeformed (though initially curved) geometry. Again see Appendix I for a rigorous derivation.

The modal Equations, (10), are modified by added terms as follows. On the right-hand side of (10a)

$$\sum_r a_r L^F_{rn}$$

and on the right-hand side of (10b)

$$\sum_m q_m L^{z_p}_{ms}$$

where

$$L^F_{rn} \equiv \iint \left[\frac{\partial^2 \phi_r}{\partial y^2}\,\frac{\partial^2 z_p}{\partial x^2} + \frac{\partial^2 \phi_r}{\partial x^2}\,\frac{\partial^2 z_p}{\partial y^2} - 2\,\frac{\partial^2 \phi_r}{\partial x \partial y}\,\frac{\partial^2 z_p}{\partial x \partial y}\right]\psi_n \,\mathrm{d}x\,\mathrm{d}y$$

$$L^{z_p}_{ms} \equiv \iint \left[2\,\frac{\partial^2 z_p}{\partial x \partial y}\,\frac{\partial^2 \psi_m}{\partial x \partial y} - \frac{\partial^2 z_p}{\partial x^2}\,\frac{\partial^2 \psi_m}{\partial y^2} - \frac{\partial^2 z_p}{\partial y^2}\,\frac{\partial^2 \psi_m}{\partial x^2}\right]\phi_s \,\mathrm{d}x\,\mathrm{d}y$$

Within a certain approximation the foregoing (which is a shallow shell or Donnell type shell theory) may even be suitable for cylindrical shells. See [43] and [44]. An extension of the basic approximation is also available for the linear theory of conical shells, see [45].

Finally we should emphasize that nominally flat plates may frequently deviate from flatness by geometrical imperfections on the order of the plate thickness. This will be of importance under some circumstances and clearly if the geometry of the imperfections can be determined by measurement, their effect will be adequately described by the shallow curved plate theory.

The effects of curvature are rather subtle. For streamwise curvature the dynamic pressure at flutter decreases, for spanwise curvature it decreases for stiff in-plane restraint and increases for flexible in-plane restraint! (See [48].) Geometric imperfections or very small curvature is particularly important for buckled plates whose flutter dynamic pressure under certain conditions would otherwise be zero.

A formal Rayleigh-Ritz approach

For the purpose of generalizing our earlier ideas we again shall employ an energy approach with Hamilton's Principle. Considering a shallow curved plate for definiteness (the same basic considerations would apply for complete cylindrical or conical shells) we may write the elastic potential energy as

$$U = U(u, v, w; z_p) \tag{17}$$

where the explicit form of (17) is given by Equations (3) and (9) of Appendix I. The kinetic energy and pressure distribution virtual work is unchanged from the linear model. We shall proceed directly to a Rayleigh-Ritz approach. Expanding the three displacements

$$\begin{aligned} w &= \sum_m q_m \psi_m \\ u &= \sum_i a_i u_i \\ v &= \sum_k b_k v_k \end{aligned} \tag{18}$$

Substituting (18) into (17) we have

$$\begin{aligned} U &= U(q_m, a_i, b_k; z_p) \\ T &= T(\dot{q}_m) \end{aligned} \tag{19}$$

The Lagrange's Equations are

$$\frac{\mathrm{d}}{\mathrm{d}t}\frac{\partial L}{\partial \dot{q}_m} - \frac{\partial L}{\partial q_m} + Q_m = 0 \tag{20a}$$

$$\frac{\partial L}{\partial a_i} = 0 \tag{20b}$$

$$\frac{\partial L}{\partial b_k} = 0 \tag{20c}$$

where

$$L \equiv T - U$$

Equations (20b) and (20c) are algebraic relations which may be solved (simultaneously) for the a_i and b_k in terms of q_m and substituted into (20a). (20a) is a system of nonlinear ordinary differential equations for the q_m.

Note that no particular support conditions have been specified. Any support conditions can be handled. In particular, if we select modes for a free-free plate (e.g., products of free-free beam natural modes) then supports may be accounted for by additional appropriate elastic energy contributions or through constraint conditions using Lagrange multipliers as in Chapter 2.

For further study the reader may consult the following sources. Fung and, subsequently, Fralich [49] examined the linear dynamic stability of a static (nonlinear) buckled equilibrium state. Carter and Stearman [50] and Barr and Stearman [51] have considered pressurized and/or buckled shells using a similar analysis. Bolotin [2] and Kobayashi [52] initiated work on dynamic nonlinear analyses of flat plates. This work was later extended in various ways including improved accuracy through the use of full potential flow aerodynamics [53] and the effects of curvature [20, 48]. Evensen and Olson [54] treated the cylindrical shell in a similar analysis. All of the above used modal methods and, with the exception of [53b], piston theory aerodynamics. Some authors have used the method of harmonic balance to treat temporal variations, others have numerically computed time histories of response. Most recently Morino, Kuo and Dugundji [55] have employed a two-time scale perturbation method which appears to have promise.

4

Aerodynamic theory

4.1 Potential flow aerodynamics

There are essentially three levels of approximation to the motion dependent aerodynamic pressures on an oscillating panel presently available. The simplest of these and consequently the most widely used is the so-called 'piston theory' which gives the following simple relation between pressure and motion

$$\Delta p^M = \rho_\infty a_\infty \left[\frac{\partial w}{\partial t} + U_\infty \frac{\partial w}{\partial x} \right] \tag{1}$$

In the above and in the following we assume that the deflection is zero at the panel leading and trailing edges. This simplifying assumption is convenient and useful but not essential [56]. The needed generalized aerodynamic forces are simply evaluated as

$$\begin{aligned} Q_n &\equiv \iint \frac{\Delta p^M}{\rho_\infty U_\infty^2} \psi_n \mathrm{d}x\,\mathrm{d}y \\ &= \frac{1}{M} \sum_m \iint \left[\frac{\dot{q}_m}{U_\infty} \psi_m + q_m \frac{\partial \psi_m}{\partial x} \right] \psi_n \mathrm{d}x\,\mathrm{d}y \end{aligned}$$

This is the first level of approximation. A more accurate theory would account for the effects of deformation at other points and at earlier times on the pressure, $p(x, t)$. Such a theory has been developed within the linear, isentropic, inviscid potential flow approximation [3, 29, 56] when the governing field equation is

$$\nabla^2\phi - \frac{1}{a_\infty^2}\left[\frac{\partial}{\partial t} + U_\infty \frac{\partial}{\partial x}\right]^2 \phi = 0 \tag{2}$$

and where ϕ is the (perturbation) velocity potential. The pressure is related to velocity potential through Bernoulli's equation

$$\Delta p^M = -\rho_\infty \left[\frac{\partial}{\partial t} + U_\infty \frac{\partial}{\partial x}\right]\phi \tag{3}$$

(2) must be solved for ϕ subject to appropriate boundary conditions. In the linear approximation for a flat plate, these are

$$\left.\frac{\partial \phi}{\partial z}\right|_{z=0} = \left[\frac{\partial}{\partial t} + U_\infty \frac{\partial}{\partial x}\right] w \tag{4}$$

and also a finiteness or outward radiation condition for $z \to \infty$ as appropriate. We shall derive Equations (2)–(4) later in the context of a more general formulation, Section 4.2. For the moment, assuming their validity let us consider how they may be used to determine the aerodynamic generalized forces.

The most economical procedure [56] is to take a Fourier Transform with respect to x and y and a Laplace Transform with respect to t. Define the transform as

$$\phi^*(z; p, \alpha, \gamma) = \int_0^\infty \iint_{-\infty}^{\infty} \phi(z, t, x, y) \mathrm{e}^{-pt - \mathrm{i}\alpha x - \mathrm{i}\gamma y} \mathrm{d}x \mathrm{d}y \tag{5}$$

Equations (2) and (4) become

$$\frac{\mathrm{d}^2\phi^*}{\mathrm{d}z^2} = \zeta^2 \phi^* \tag{6}$$

and

$$\left.\frac{\mathrm{d}\phi^*}{\mathrm{d}z}\right|_{z=0} = \left[\frac{\partial w}{\partial t} + U_\infty \frac{\partial w}{\partial x}\right]^* \equiv f^* \tag{7}$$

where

$$\zeta^2 \equiv \frac{p^2}{a_\infty^2} + 2M\frac{pi\alpha}{a_\infty} - \alpha^2(M^2 - 1) + \gamma^2$$

Solving (6), we have

$$\phi^* = A\mathrm{e}^{+\zeta z} + B\mathrm{e}^{-\zeta z} \tag{8}$$

Formally dropping A because of the finiteness condition for $z \to \infty$ and evaluating B using (7), we obtain

$$\phi^* = -\frac{f^*}{\zeta}\mathrm{e}^{-\zeta z} \tag{9}$$

Inverting (9) with respect to time

$$\mathscr{L}^{-1}\phi^*|_{z=0} =$$

$$-a_\infty \int_0^t \mathscr{L}^{-1}f^*(t)\exp[-\mathrm{i}U_\infty\alpha(t-\tau)]J_0[(\alpha^2+\gamma^2)^{\frac{1}{2}}a_\infty(t-\tau)]\,\mathrm{d}\tau \tag{10}$$

Using a transformed Bernoulli's equation, (3), and (10), we may compute the transform of the pressure.

$$\mathscr{L}^{-1}\Delta p^{M*} = \rho_\infty a_\infty\Big\{\mathscr{L}^{-1}f^*(t) - a_\infty(\alpha^2+\gamma^2)^{\frac{1}{2}}\cdot$$

$$\cdot\int_0^t \mathscr{L}^{-1}f^*(\tau)\exp[-\mathrm{i}U_\infty\alpha(t-\tau)]J_1[a_\infty(\alpha^2+\gamma^2)^{\frac{1}{2}}(t-\tau)]\,\mathrm{d}\tau\Big\} \tag{11}$$

From this we see the well-known result [3] that at $t=0^+$, the pressure is that given by 'piston theory'. Formally we may invert to the spatial domain to obtain a relation like (2.3). However, since our primary purpose is to evaluate the aerodynamic generalized forces we may more efficiently evaluate these directly using the definition

$$Q_{mn} \equiv \iint \frac{\Delta p_m}{\rho_\infty U_\infty^2}\psi_n\,\mathrm{d}x\,\mathrm{d}y \tag{12}$$

where Δp_m is pressure due to $w = q_m \psi_m$, and first integrating with respect to x and y before inverting with respect to α and γ. The result is

$$Q_{mn} = q_m(t) S_{mn} + \dot{q}_m(t) D_{mn} + \int_0^t [q_m(\tau) H_{mn}(t-\tau) + \dot{q}_m(\tau) I_{mn}(t-\tau)] \mathrm{d}\tau \tag{13}$$

where

$$S_{mn} \equiv \frac{1}{M} \iint \frac{\mathrm{d}\psi_m}{\mathrm{d}x} \psi_n \mathrm{d}x \mathrm{d}y$$
$$D_{mn} \equiv \frac{1}{M} \frac{1}{U_\infty} \iint \psi_m \psi_n \mathrm{d}x \mathrm{d}y \tag{13a}$$

$$H_{mn}(t) = -\frac{1}{4\pi^2} \frac{U_\infty}{M^2} \iint_{-\infty}^{\infty} G_{mn} \mathrm{i}\alpha [\alpha^2 + \gamma^2]^{\frac{1}{2}} \cdot \exp[-\mathrm{i}\alpha U_\infty t] J_1[a_\infty(\alpha^2 + \gamma^2)^{\frac{1}{2}} t] \mathrm{d}\alpha \mathrm{d}\gamma \tag{13b}$$

$$I_{mn}(t) = -\frac{1}{4\pi^2} \frac{1}{M^2} \iint_{-\infty}^{\infty} G_{mn} [\alpha^2 + \gamma^2]^{\frac{1}{2}} \cdot \exp[-\mathrm{i}\alpha U_\infty t] J_1[a_\infty(\alpha^2 + \gamma^2)^{\frac{1}{2}} t] \mathrm{d}\alpha \mathrm{d}\gamma$$

$$G_{mn} \equiv \iint \psi_m(x, y) \mathrm{e}^{-\mathrm{i}\alpha x - \mathrm{i}\gamma y} \mathrm{d}x \mathrm{d}y \cdot \iint \psi_n(x, y) \mathrm{e}^{+\mathrm{i}\alpha x + \mathrm{i}\gamma y} \mathrm{d}x \mathrm{d}y \tag{13c}$$

The integrals in (13b) must be performed numerically. Those in (13a) and (13c) may be done analytically for most panel modal expansions. Note that although we have chosen to use rectangular cartesian coordinates there is no restriction on panel shape in the analysis and, in particular, the integrals in (13c) are over the panel area whatever its shape. A similar calculation has been done for the cylindrical shell geometry [57]. Also see [46] and [47].

The above constitutes the second level of approximation. An alternative numerical approach is the so-called 'Mach box' method. See [3] and [29]. This is essentially a finite difference approximation to the integral equation solution of the wave equation and is usually a less efficient method for the present problem.

Aerodynamic pressure forms and approximations

For reasons already discussed, when employing the more elaborate aerodynamic theories it is efficient to bypass the calculation of pressure and deal directly with generalized forces. However, one can gain some qualitative insight into the possible aeroelastic behavior by considering the aerodynamic pressure and various approximations thereto, as we have seen in Chapter 2.

The approximate aerodynamic pressure relations used there will now be derived. We shall deal with a two-dimensional flow over a flat plate but similar results hold for three-dimensional flow including those over cylindrical shells. Moreover we shall avoid the interesting difficulties associated with possible spatial discontinuities at the leading and trailing edges, by extending our integrations over all space.

From Equation (11) we may calculate the aerodynamic pressure

$$\frac{\Delta p^M}{\rho_\infty U_\infty^2} = \frac{1}{M}$$

$$\cdot \left\{ \frac{f}{U_\infty} + \int_{-\infty}^{\infty} \int_0^t A(x - \xi, t - \tau) f\left(\frac{\xi}{a}, \frac{\tau U_\infty}{a}\right) \right\} d\frac{\tau U_\infty}{a} \frac{d\xi}{a} \tag{14}$$

where

$$A\left(\frac{x}{a}, \frac{tU_\infty}{a}\right) = -\frac{1}{2\pi M} \int_{-\infty}^{\infty} \alpha^* e^{-i\alpha^* \frac{tU_\infty}{a}} J_1 \left[\frac{\alpha^* \frac{tU_\infty}{a}}{M}\right] e^{i\alpha^* x/a} d\alpha^*$$

As already noted the first term is that associated with so called piston theory aerodynamics. For $M \gg 1$, we see that the integral term is of order $1/M^2$ compared to the first term and hence this allows an assessment of the error involved in using piston theory aerodynamics.

If one assumes sinusoidal motion, i.e.,

$$w = \bar{w}e^{i\omega t}$$
$$f = \bar{f}e^{i\omega t}$$
$$\Delta p = \Delta\bar{p}e^{i\omega t}$$

and expands the result in a power series in ω, one determines for supersonic flow that

$$\frac{\Delta\bar{p}^M}{\rho_\infty U_\infty^2} = \frac{1}{(M^2-1)^{\frac{1}{2}}}\left\{\frac{\partial\bar{w}}{\partial x} + \frac{M^2-2}{M^2-1}\frac{i\omega}{U_\infty}\bar{w}\right\} + 0(\omega^2) \tag{15}$$

For small reduced frequencies retention of only the first two terms in the series should be a suitable approximation. This series converges more rapidly as Mach number increases as one would expect. Compare 4.1-15 and 2.1-30.

Another interesting form may be obtained by computing the transformed pressure before inverting the velocity potential with respect to time. The result is

$$\frac{\Delta p^M}{\rho_\infty U_\infty^2} = \int_0^t \frac{U_\infty}{a}\int_{-\infty}^{\infty} A^{SB}(\xi, t-\tau)$$
$$\cdot\left[\frac{\partial}{\partial\tau U_\infty/a} + \frac{\partial}{\partial(x-\xi)/a}\right]^2 \frac{w(x-\xi)}{a}\frac{d\xi}{a}\frac{d\tau U_\infty}{a} \tag{16}$$

where $p \equiv i\omega$

$$A^{SB} \equiv \frac{1}{(2\pi)^2}\iint_{-\infty}^{\infty} \frac{e^{+i\alpha^* x/a + i\omega t}}{\zeta a}\, d\frac{\omega a}{U_\infty}\, d\alpha^*$$

and

$$\zeta a \equiv \left[\left(\frac{pa}{U_\infty}\right)^2 M^2 + 2iM^2\alpha^*\frac{pa}{U_\infty} - \alpha^{*2}(M^2-1)\right]^{\frac{1}{2}}$$

$$\alpha^* \equiv \alpha a$$

This form is most significant for $M \ll 1$. For $M \equiv 0$, $\zeta a = |\alpha^*|$. Hence

$$A^{SB} = \frac{\delta(t)}{2\pi} \int_{-\infty}^{\infty} \frac{e^{i\alpha^* x/a}}{|\alpha^*|} \, d\alpha^* = \frac{-\delta(t)}{\pi} \ln x/a$$

A form which is sometimes referred to as the 'slender-body approximation' can be obtained by expanding w about x (for simplicity assuming $w(0) = w(a) = 0$) and integrating. The result is

$$\frac{\Delta p^M}{\rho_\infty U_\infty^2} = -\sum_i A_i \frac{\partial^{i+2}(w/a)}{\partial(x/a)^{i+2}}, \quad A_i \equiv \frac{1}{i!\pi} \int_{-x/a}^{1-x/a} \eta \ln |\eta| \, d\eta \tag{17}$$

If one truncates the series to the first term, then one obtains Equation 2.1-29.

This approximation is not quantitatively accurate because of the poor convergence of the series. However for a cylindrical shell of large length/radius such an approximation may even be quantitatively useful. See [57] for details.

We have given results using the above aerodynamic approximations previously in Chapter 2. Hence we shall proceed directly to the most accurate aerodynamic approximation presently available before presenting additional results.

4.2 Shear flow aerodynamics

For sufficiently thick aerodynamic boundary layers the inviscid, potential flow model becomes inadequate. The boundary layer effect is particularly important at transonic-low supersonic Mach numbers where, for example, a boundary layer thickness of one-tenth the panel chord may result in an increase in flutter dynamic pressure of several hundred percent. To account for the fluid boundary layer the third and most accurate level of approximation has been developed [58].

To construct a model which will acount for the principal boundary layer effects requires some physical assumptions about the flow. We shall assume that the principal effects are due to the variation in the mean velocity (and temperature) profiles within the boundary layer and that one may ignore the direct effect of viscosity on the time dependent pressure due to panel motion, i.e., a 'shear flow' model will be employed. This would seem reasonable and consistent with the assumption that the plate motion does not modify the turbulent pressure fluctuations. The mean motion itself will be taken as prescribed from experiment or an appropriate viscous flow theory. We shall develop a mathematical analysis for the pressure fluctuations due to panel motion about this prescribed state.

To derive an equation of motion for the pressure perturbations it is sufficient to consider nonlinear, inviscid flow. The continuity of mass equation reads

$$\frac{\partial \rho}{\partial t} + \frac{\partial(\rho u)}{\partial x} + \frac{\partial(\rho v)}{\partial y} + \frac{\partial(\rho w)}{\partial z} = 0 \tag{1}$$

The three momentum equations read:

$$\begin{aligned} \rho \frac{Du}{Dt} &= - \frac{\partial p}{\partial x} \\ \rho \frac{Dv}{Dt} &= - \frac{\partial p}{\partial y} \\ \rho \frac{Dw}{Dt} &= - \frac{\partial p}{\partial z} \end{aligned} \tag{2}$$

$$\frac{D}{Dt} \equiv \frac{\partial}{\partial t} + u \frac{\partial}{\partial x} + v \frac{\partial}{\partial y} + w \frac{\partial}{\partial z}$$

The energy equation reads:

$$\rho \left\{ c_v \frac{DT}{Dt} + \frac{D}{Dt} \left(\frac{u^2 + v^2 + w^2}{2} \right) \right\}$$

$$= - \frac{\partial}{\partial x} (pu) - \frac{\partial}{\partial y} (pv) - \frac{\partial}{\partial z} (pw) \tag{3}$$

Finally we have the perfect gas formula

$$p = \rho RT \tag{4}$$

We shall assume that the various quantities may be written as the sum of a mean plus perturbation.

$$\begin{aligned}
u &= \bar{u}(z) + \hat{u} \\
v &= \hat{v} \\
w &= \hat{w} \\
p &= \bar{p}(z) + \hat{p} \\
\rho &= \bar{\rho}(z) + \hat{\rho} \\
T &= \bar{T}(z) + \hat{T}
\end{aligned}$$

The subsequent analysis may be carried through for any given mean flow; however, for our purposes here we use a boundary layer type mean flow. That is we shall assume that

$$\bar{p} = p_\infty \text{ for all } z$$

and thus

$$\bar{\rho}\bar{T} = \rho_\infty T_\infty \tag{6}$$

and finally from conservation of energy

$$\bar{T}/T_\infty = 1 + \left(\frac{\gamma - 1}{2}\right) M_\infty^2 \left[1 - \left(\frac{\bar{u}}{U_\infty}\right)\right]^2 \tag{7}$$

From the last equation we may determine $\bar{T}$ once $\bar{u}$ is prescribed and then compute $\bar{\rho}$ from (6).

The equation for $\hat{p}$ may be obtained by substituting (5) into (1) through (4) and linearizing the resulting equations in the perturbation quantities. By suitable cross-differentiation and elimination a single equation for $\hat{p}$ is obtained.

$$\frac{M_\infty^2}{\bar{T}/T_\infty} D^3\hat{p} - D\nabla^2\hat{p}U_\infty^2 + 2\frac{\partial^2\hat{p}}{\partial x \partial z}\frac{\mathrm{d}\bar{u}}{\mathrm{d}z}U_\infty^3 - D\frac{\partial\hat{p}}{\partial z}\frac{1}{T}\frac{\mathrm{d}\bar{T}}{\mathrm{d}z}U_\infty^2 = 0 \tag{8}$$

where

$$D \equiv \frac{\partial}{\partial t} + \bar{u}\frac{\partial}{\partial x}$$

The boundary condition at the plate surface is

$$\left.\frac{\partial \hat{p}}{\partial z}\right|_{z=0} = -\bar{\rho}\left[\frac{\partial}{\partial t} + \bar{u}\frac{\partial}{\partial x}\right]^2 w_p\bigg|_{z=0} \tag{9}$$

Equation (9) is derivable using the usual velocity boundary condition

$$\hat{w}\bigg|_{z=0} = \left(\frac{\partial}{\partial t} + \bar{u}\frac{\partial}{\partial x}\right) w_p\bigg|_{z=0} \tag{10}$$

and the (perturbation) z momentum equation. Because $\bar{u} \to 0$ as $z \to 0$ it is not clear a priori that one may accurately evaluate the boundary condition on $z = 0$ rather than $z = w_p$. Results have shown this is a reasonable approximation for at least the range of parameters investigated to date.

We shall now turn to the solution of (8) and the evaluation of the generalized aerodynamic forces. The basic procedure is similar to that employed for the potential flow analysis. Fourier transforms are employed in x and y to reduce (8) to a partial differential equation in z, t, i.e., defining

$$p^* \equiv \iint_{-\infty}^{\infty} \hat{p}\, e^{-i\alpha x - i\gamma y}\, dx\, dy \tag{11}$$

(8) becomes an equation for p^* with parameters α and γ. This partial differential equation is further reduced to a system of ordinary differential equations in time using a finite difference representation in z from the wall $z = 0$ to some distant wall $z = z_{max}$. The values of $\bar{u}$, $\bar{T}$ employed are those at the mid-point of each difference element. For brevity we do not write out this system of equations in detail; the technique is standard. This system of equations is solved using step-by-step integration in time for a step function change in (the Fourier Transform of)

$$\frac{\bar{\rho}}{\rho_\infty}\left[\frac{\partial}{\partial t} + \bar{u}\frac{\partial}{\partial x}\right]^2 w_p\bigg|_{z=0}$$

to obtain the (Fourier Transform of) the pressure at the wall, $p_w = p_{z=0}$ Call this solution K; then the solution for any arbitrary time dependent motion can be written

$$\frac{p_w^*}{\rho_\infty}(t; \alpha, \gamma) = K'(t; \alpha, \gamma)\left[\frac{\partial}{\partial t} + \mathrm{i}\alpha u_w\right] w_p^*(t; \alpha, \gamma)\Big|_{t=0}$$

$$+ w_p^*(t; \alpha, \gamma)\Big|_{t=0}\left[\frac{\partial}{\partial t} + \mathrm{i}\alpha u_w\right] K'(t; \alpha, \gamma)$$

$$+ \int_0^t K'(t-\tau; \alpha, \gamma)\left[\frac{\partial}{\partial \tau} + \mathrm{i}\alpha u_w\right]^2 w_p^*(\tau; \alpha, \gamma)\,\mathrm{d}\tau \qquad (12)$$

where

$$K' \equiv \frac{\mathrm{d}K}{\mathrm{d}t}$$

and

$$w_p^* \equiv \iint_{-\infty}^{\infty} w_p \mathrm{e}^{-\mathrm{i}\alpha x - \mathrm{i}\gamma y}\,\mathrm{d}x\,\mathrm{d}y \qquad (13)$$

Formally (12) may be inverted to obtain

$$p_w = \frac{1}{(2\pi)^2}\iint_{-\infty}^{\infty} p_w^* \mathrm{e}^{+\mathrm{i}\alpha x + \mathrm{i}\gamma y}\,\mathrm{d}x\,\mathrm{d}y \qquad (14)$$

but again it is more efficient to deal directly with the generalized forces

$$Q_{mn} = \iint \frac{\Delta p_m^M}{\rho_\infty U_\infty^2}\psi_n\,\mathrm{d}x\,\mathrm{d}y$$

$$\Delta p^M \equiv p_w \qquad (15)$$

Using (13) and (14) in (15) we obtain

$$Q_{mn} = q_{mn}(0)[H'_{mn}(t) + 2I_{mn}(t)]$$

$$+ \dot{q}_{mn}(0)H_{mn}(t)$$

$$+\int_0^t \{H_{mn}(t-\tau)\ddot{q}_m(\tau)$$

$$+ 2I_{mn}(t-\tau)\dot{q}_m(\tau)$$

$$+ K_{mn}(t-\tau)q_m(\tau)\}\,\mathrm{d}\tau \tag{16}$$

where

$$\begin{aligned}
H_{mn} &\equiv \left(\frac{1}{2\pi}\right)^2 \iint_{-\infty}^{\infty} \frac{K'}{U_\infty^2} G_{mn}\,\mathrm{d}\alpha\,\mathrm{d}\gamma \\
I_{mn} &\equiv \left(\frac{1}{2\pi}\right)^2 \bar{u}_w \iint_{-\infty}^{\infty} \frac{K'\mathrm{i}\alpha}{U_\infty^2} G_{mn}\,\mathrm{d}\alpha\,\mathrm{d}\gamma \\
K_{mn} &\equiv \left(\frac{1}{2\pi}\right)^2 \bar{u}_w^2 \iint_{-\infty}^{\infty} \frac{K'}{U_\infty^2} (\mathrm{i}\alpha)^2 G_{mn}\,\mathrm{d}\alpha\,\mathrm{d}\gamma
\end{aligned} \tag{17}$$

and G_{mn} is as defined in (4.1-13c). Finally we note that an analytical formula is available for K' in the inviscid potential flow limit

$$K'(t;\alpha,\gamma) = a_\infty \mathrm{e}^{-\mathrm{i}\alpha U_\infty t} J_0[(\alpha^2+\gamma^2)^{\frac{1}{2}} t a_\infty] \tag{18}$$

With regard to the inviscid, potential flow approximation this is obtained by setting

$$\begin{aligned}
\bar{u} &= \text{constant} = U_\infty \\
\bar{T} &= \qquad\qquad\quad T_\infty
\end{aligned} \tag{19}$$

in the above relations and recalling the definition of velocity potential

$$\hat{q} = \nabla\phi \tag{20}$$

All perturbation quantities then satisfy, see (8)

$$M_\infty^2 D^2 \hat{p} - \nabla^2 \hat{p} U_\infty^2 = 0 \tag{21}$$

which is just (4.1-2) written in a different form.

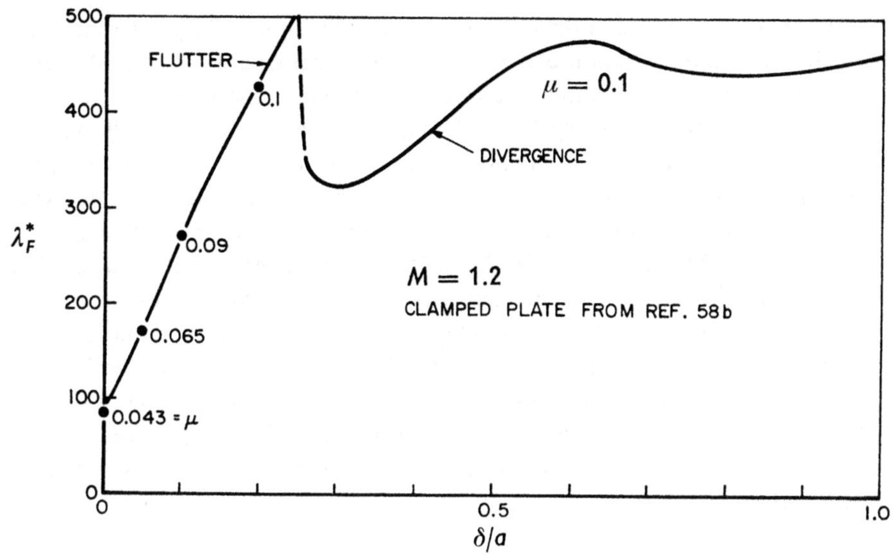

Fig. 4.1 Flutter dynamic pressure vs. boundary layer thickness.

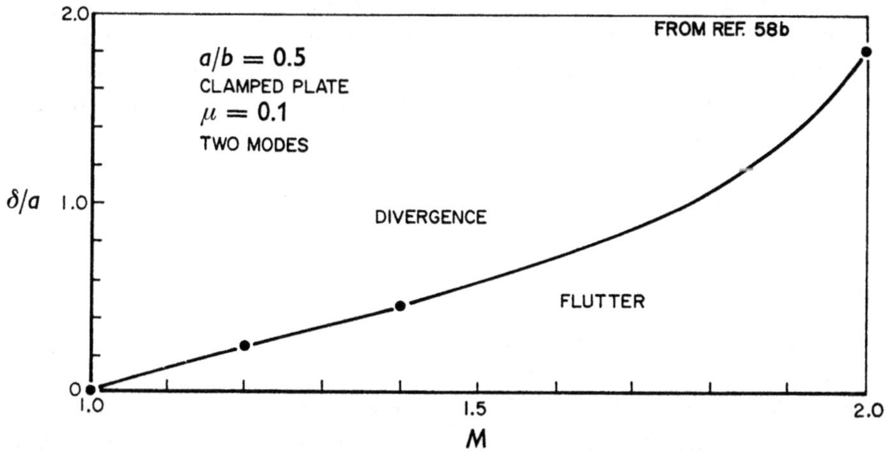

Fig. 4.2 Boundary layer thickness required for suppressing flutter.

It has been found that the boundary layer effect normally gives an increase in flutter dynamic pressure (based on free stream parameters). Moreover one of the most interesting results of the aerodynamic analysis is that for sufficiently large boundary layer thickness, a panel will never flutter but only diverge. In Fig. 4.1 typical results are shown for flutter (or divergence) dynamic pressure *vs.* (non-dimensional) boundary layer thickness, δ/a, while in Fig. 4.2 the boundary thickness which separates flutter from divergence is plotted *vs.* Mach number. In the latter figure we see that the δ/a required to suppress flutter entirely increases rapidly with M. All of these results are associated with the effective decrease in flow velocity, dynamic pressure and Mach number (relative to nominal free stream values) created by the presence of the boundary layer. In simple terms, one may consider the foregoing aerodynamic analysis as a means of calculating these effective flow parameters. Single mode flutter is particularly sensitive to boundary layer effects.

4.3 Angle of flow effects

For a given panel geometry, say a rectangular panel, the flow may be yawed in the plane of the panel rather than aligned with a panel edge. Formally no modification need be made to the above analyses, however it may be convenient to do the aerodynamic analysis in a coordinate fixed with respect to the structure rather than the flow. For example, for a rectangular panel with coordinates x', y' and flow coordinates x, y we simply make a coordinate transformation

$$\frac{\partial}{\partial x} = \frac{\partial}{\partial x'}\frac{\partial x'}{\partial x} + \frac{\partial}{\partial y'}\frac{\partial y'}{\partial x}$$

$$\frac{\partial}{\partial y} = \frac{\partial}{\partial x'}\frac{\partial x'}{\partial y} + \frac{\partial}{\partial y'}\frac{\partial y'}{\partial y}$$

where derivatives such as $\partial x'/\partial x$, etc., may be evaluated from elementary geometry. Doing the aerodynamic problem in the x', y' coordinates offers only a modest algebraic complication to the basic analysis if one uses a Fourier Transform approach.

Inclination of the panel to the plane of the oncoming flow can also be important under some circumstances. Local steady flow conditions over the panel, substantially different from those in the free stream, may be significant for panel flutter. It has generally been believed that the use of the local flow pressure, density and Mach number, in place of the free stream values in the aerodynamic theory, was a satisfactory approximation for the fluid forces. A study by Bailie and McFeely [59] has shown this to be true for a panel mounted on a wedge in hypersonic flow. Their results, using a full unsteady hypersonic theory, which accounts for the change in the steady flow field, agree very closely with the simpler more approximate theory suggested above.

The principal effect of hypersonic flow on the flutter results is that the flutter dynamic pressure does not increase linearly with Mach number for large M (as predicted by unmodified piston theory) but asymptotes instead to some constant value depending on the wedge angle (or, more generally, the degree of bluntness of the body on which the panel is mounted). The lowest flutter dynamic pressures, however, still occur at low supersonic-transonic free stream Mach number. Therefore, it would appear that, unless the body is extremely blunt and/or traverses a trajectory with maximum free stream dynamic pressure at hypersonic Mach number, the hypersonic regime will not be the most critical with regard to flutter.

The more general question of accounting for local flow conditions at all Mach number remains, however. For some geometries (more complicated than the wedge) it may no longer be possible to treat the mean steady flow over the panel as uniform along the plate length. Whether general analytical or even useful semi-empirical methods may be developed remains to be seen. What may well evolve, at least in the near future, are ad hoc tests for the particular configuration to determine the steady-flow field combined with the use of average flow properties in a flutter analysis.

4.4 Cavity acoustics

A cavity enclosure beneath the panel may be of significance for two reasons. Firstly the cavity dynamics may affect panel response and secondly the cavity dynamics themselves can be important from the standpoint of vibration and sound transmission. We shall use an inviscid, potential flow model where the cavity pressure must satisfy (in a volume V with surface area A)

$$\nabla^2 p_c - \frac{1}{a_c^2}\frac{\partial^2 p_c}{\partial t^2} = 0 \tag{1}$$

where

$$\begin{aligned} \frac{\partial p_c}{\partial n} &= 0 && \text{on rigid wall area } A_R \\ &= -\rho_c \frac{\partial^2 w}{\partial t^2} && \text{on flexible wall area } A_F \end{aligned} \tag{2}$$

and where n – normal to wall, with positive direction out of the fluid. w is also positive out of the fluid as is p_c.

Because we are dealing with an enclosed fluid volume we choose to describe the cavity fluid motion in terms of its (rigid wall) acoustical normal modes F_m which satisfy

$$\begin{aligned} \nabla^2 F_m &= -\left(\frac{\omega_m^c}{a_c}\right)^2 F_m \\ \frac{\partial F_m}{\partial n} &= 0 \quad \text{on} \quad A_F + A_R \end{aligned} \tag{3}$$

and an orthogonality condition

$$\begin{aligned} \frac{1}{V}\iiint F_m F_n \mathrm{d}V &= 0 && \text{for } m \neq n \\ &\equiv M_n^c && \text{for } m \neq n \end{aligned} \tag{4}$$

Note that the equation (3) has a solution $\omega_1^c = 0$, $F_1 = 1$. All other cavity natural frequencies ω_m^c, $m = 2, 3, \ldots$, are non-zero however. For a rectangular cavity with wall dimensions, a, b, d

$$\left(\frac{\omega_m^c}{a_c}\right)^2 = \pi^2\left[\left(\frac{i_m}{a}\right)^2 + \left(\frac{j_m}{b}\right)^2 + \left(\frac{k_m}{b}\right)^2\right]$$

and

$$F_m = \cos\frac{i_m \pi x}{a}\cos\frac{j_m \pi y}{b}\cos\frac{k_m \pi z}{d}$$

where i_m, etc. $= 0, 1, 2, 3, \ldots$

The wave equation can be transformed into a set of ordinary differential equations in time by using Green's Theorem in the form

$$\iiint_V (p_c \nabla^2 F_m - F_m \nabla^2 p_c)\,\mathrm{d}V = \iint\left[p_c \frac{\partial F_m}{\partial n} - F_m \frac{\partial p_c}{\partial n}\right]\mathrm{d}A \tag{5}$$

Defining

$$p_m \equiv \frac{\iiint p_c F_m \,\mathrm{d}V}{\rho_c a_c^2 V}$$
$$w_m \equiv \frac{\iint w F_m|_{A_F}\,\mathrm{d}A}{A_F} \tag{6}$$

and making use of the fact that p_c and F_m satisfy (1), (2) and (3), the following ordinary differential equations are obtained from (5)

$$\ddot{p}_m + (\omega_m^c)^2 p_m = -\ddot{w}_m \frac{A_F}{V} \tag{7}$$

From (6) we recognize the p_m as the coefficients of a normal mode expansion for the pressure

$$p_c = \rho_c a_c^2 \sum_m \frac{p_m F_m}{M_m}\bigg|_{A_F} \tag{8}$$

Expanding the flexible panel deflection as

$$w = \sum_n q_n \psi_n \tag{9}$$

we define a cavity generalized force as

$$Q_n^c \equiv \frac{\iint_{A_F} p_c \psi_n \,\mathrm{d}A}{\rho_c a_c^2} \tag{10}$$

From (6) and (9) we obtain

$$w_m = \sum_r L_{mr} q_r \tag{11}$$

where

$$L_{mr} \equiv \frac{\iint_{A_F} F_m \psi_r \, \mathrm{d}A}{A_F}$$

Thus (7) becomes

$$\ddot{p}_m + (\omega_m^c)^2 p_m = -\frac{A_F}{V} \sum_r L_{mr} \ddot{q}_r \tag{12}$$

From (8) and (10) we obtain

$$Q_n^c = A_F \sum \frac{p_m L_{mn}}{M_m} \tag{13}$$

Our solution procedure then is to solve for p_m from (12) and then calculate Q_n^c from (13). The former may be determined either analytically in the form of a convolution integral or, more conveniently from the standpoint of the total problem, step-by-step in time. Please note that with our sign convention

$$\rho_c a_c^2 Q_n^c \delta q_n$$

represents positive work. Hence the total aerodynamic or acoustic generalized force is

$$\rho_\infty U_\infty^2 Q_n - Q_n^c \rho_c a_c^2$$

which will appear in this combination in all of our structural equations of motion.

The above analysis is due to C. S. Ventres (unpublished) although similar work has been reported in the acoustical literature [60].

If one is interested in the cavity pressure field for its own sake, e.g., the sound transmitted from the boundary layer turbulent pressure fluctuations through the structure into the cavity, the above complete analysis is needed. However, as regards to the panel motion per se, the cavity acoustic pressures will significantly modify the panel response only for shallow cavity depths, $d/a \ll 1$. For such shallow cavities a simpler approximation will usually suffice. That is the cavity can be treated as a Helmholtz resonator and only the lowest cavity acoustic mode need be used, $\omega_1 = 0$, $F_1 = 1$. This can be shown to be equivalent to a cavity perturbation pressure given by

$$p_c = -\rho_c a_c^2 \frac{\iint_{A_F} w \, \mathrm{d}x \, \mathrm{d}y}{V} \tag{14}$$

This is an equivalent acoustic spring which effectively stiffens the plate.

The above cavity analysis can be combined with an appropriate structural theory to determine the change in panel natural modes and frequencies with cavity depth (as well as the effect on flutter). In Fig. 4.3 such results are given for the lowest two symmetric panel modes. An interesting result is that as the cavity depth decreases the two natural frequency branches interchange their modal character with the first mode branch becoming the

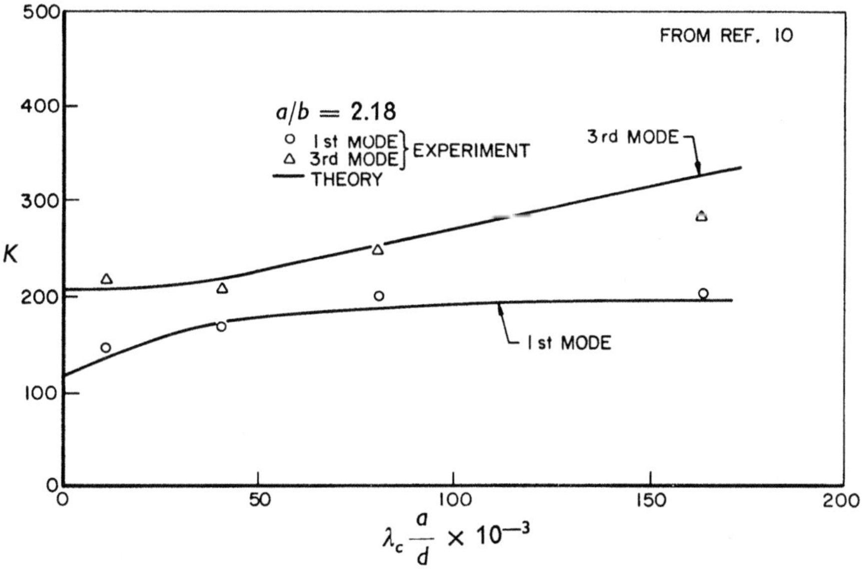

Fig. 4.3 *Effect of cavity on panel natural frequencies.*

second and vice versa. For even smaller cavity depths there are similar interchanges in turn with each of the higher symmetrical modes. The anti-symmetrical modes are relatively unchanged by all this with their frequencies decreasing slightly due to virtual mass effects which are neglected in the Helmholtz resonator approximation. For a dense fluid, e.g., water, the latter could be more significant. See [61] and [62] for further details and [10] for typical flutter results.

5

Structural response to turbulent boundary layer pressure fluctuations

5.1 Physical nature of the problem

The vibration of a flexible structure immersed in a fluid flow is again considered with particular attention now devoted to the relationship of externally excited response (due to turbulent boundary layer pressure fluctuations, 'noise') to self-excited response (due to aeroelastic instability, 'flutter').

To recapitulate, in its basic physical ingredients the problem is as follows: A thin plate or shell is immersed in a fluid stream (see Fig. 2.1). Over the upper surface of the structure there is a parellel fluid flow; beneath or inside there is a fluid-filled cavity. Here we wish to answer two basic questions about this fluid-solid (aeroelastic) system: What is the motion of the structure?; and what is the sound (fluctuating pressure) level in the cavity? The basic elements in a theoretical analysis to answer these questions may be indicated by a block diagram as in Fig. 5.1. We assume that the aerodynamic pressure on the upper surface of the structure may be divided into two parts; one due to the random turbulent nature of the flow which is independent of the structural motion and the second due to the structural motion itself. The structural motion gives rise to cavity fluid motion and, in turn, the cavity fluid pressure modifies the structural motion.

In the past, the total program indicated by Fig. 5.1 has rarely been carried out. Instead two distinct paths have been followed which we shall call here 'conventional noise analysis' and 'conventional flutter analysis', respectively. In the former the path is 'cut' as indicated in Fig. 5.1. That is, the effect of

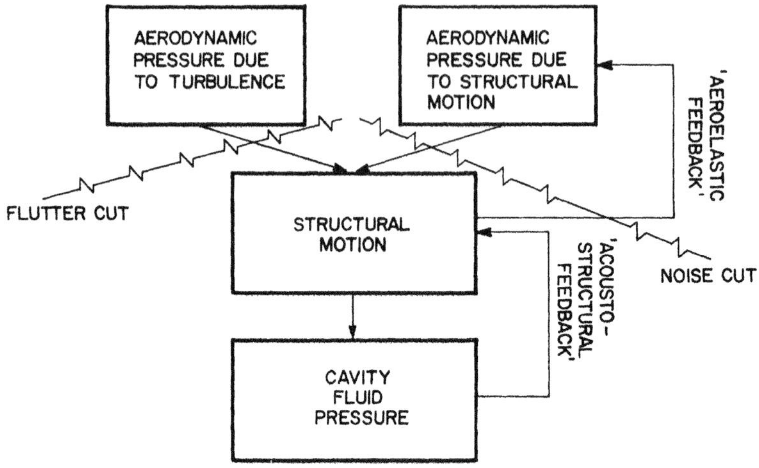

Fig. 5.1 Block diagram of analysis.

the aerodynamic pressure due to structural motion on the structural motion is neglected. On the other hand, the path is cut at a different point in a 'conventional flutter analysis'. That is, the external pressure due to turbulence is neglected. The 'conventional flutter analysis' has formed the subject of Chapters 1–4.

Numerous investigators have provided 'conventional analyses' for structural response to 'noise'. Most, though not all, have also made the following assumptions:

(1) the turbulent pressure field is assumed known from some independent source (usually experiment);
(2) the structural motion due to the turbulence is assumed sufficiently small so that linear structural theory may be employed;
(3) the acoustic cavity pressure field excited by the structural motion is assumed to have no significant effect on the plate motion. Hence not only aeroelastic but also acousto-structural feedback is neglected. See Fig. 5.1.

Using the above assumptions, the mathematical apparatus of the linear theory of stationary, ergodic, random processes (power spectral method) has been employed to obtain solutions.

A recent example of this type of analysis is the work by Strawderman

[63, 64] who discusses much of the earlier work. Also see the review by Bies [65] and the discussion in Lin [27].

Various authors have appreciated the deficiencies for some applications of using linear structural theory (which generally assumes the structure's deflection is small compared to its thickness), and neglect of aerodynamic or acoustic pressures due to the structural motion on the structural motion.

A more complete analysis would include nonlinear structural theory, and *mutual* interaction between the structure and the air flow over its upper surface and also the air in the cavity beneath. It is the purpose of this chapter to outline such an analysis. Much of the analysis [66, 67] makes use of techniques previously developed for the flutter (stability) problem of a plate or shell structure in a fluid flow where inclusion of structural nonlinearities and mutual structure-flow interaction are essential to a correct description of the physical phenomenon.

One other 'novelty' of the present analysis relative to 'conventional noise analysis' is the method of solution. Numerical integration in the time domain is employed rather than using the power spectral approach in the frequency domain. Although the power spectral approach could be used when accounting for the fluid motion forces on the structure and also structural nonlinearities (through the concept of equivalent linearization), it is at best inconvenient. In general, the power spectral methods are to be preferred only if a sufficient number of approximations are made to obtain analytical solutions. For more accurate (and unfortunately more elaborate) numerical solutions, simulation in the time domain is usually to be preferred. It should be emphasized that the present approach complements the analytical power spectral approach, but does not replace it.

A possible drawback to the time domain solution is the fact that most of the raw pressure time histories of the turbulent external flow have been digested into power spectra. However, even if the time histories are not available for simulation, there are methods for generating time histories from power spectra [68, 69]. If the time history is available (e.g., from measurement) it may be used directly as input to the present analysis without any consideration of power spectra.

Another drawback of the time versus frequency domain (power spectral) approach which is sometimes cited is the greater simplicity of the aerodynamic or acoustic forces in the frequency plane. In Chapter 4 these forces have been obtained for arbitrary time-dependent motion without undue complication.

Finally, the present method may be used to investigate the structural and cavity response to any type of external force not only turbulent pressure fluctuations.

The experimental evidence and the requisite analysis

To gain further perspective on our subject recall the results of a laboratory wind tunnel test. In Fig. 1.2 a schematic composite of results from several experimental programs was presented. There the *rms* displacement of the structure, w, non-dimensionalized by its thickness, h, was plotted versus dynamic pressure, q. Also shown were schematic results from 'conventional flutter analysis' and 'conventional noise analysis'.

For small q, the plate responds to the external 'noise' and the 'aeroelastic feedback' is unimportant. For this regime a 'conventional noise analysis' gives results in reasonable agreement with experiment. Both theory and experiment indicate the oscillation takes the form of a random oscillation with predominant frequency components near one or more of the natural modes of the structure.

Above a certain value of q, $q \simeq q_{\text{flutter}}$, the structural response increases rapidly with the character of the oscillation changing from a random to a periodic oscillation. This larger amplitude motion is the nonlinear limit cycle oscillation, 'flutter', associated with an unstable aeroelastic system. That is, a linear 'conventional flutter analysis' would predict that for any $q > q_{\text{flutter}}$ the system would be unstable and the structural deflection grow exponentially with time. If the relevant (structural) nonlinearities are included in the analysis, the finite structural amplitude may be determined for $q > q_{\text{flutter}}$ (as shown in Fig. 1.2). The additional structural response to noise is frequently small by comparison to the flutter response in this range of q and thus may be neglected.

A total analysis would, of course, determine the structural behavior for all q and treat 'noise' and 'flutter' in a unified manner. There are additional reasons for a total analysis beyond the conceptual benefits of unification.

(1) There is an intermediate range near $q = q_{\text{flutter}}$ for which either 'conventional analysis' will be inadequate. The larger the response in the 'noise' regime, $q < q_{\text{flutter}}$, the broader this intermediate range.

(2) If the 'noise response' is on the order of the plate thickness then it will generally be necessary to include structural nonlinearities indepen-

dent of any considerations of flutter. This is also true for pressurized or buckled structures. A common need for including structural nonlinearities for 'noise' as well as 'flutter' suggests a common analysis. Rather distinct types of 'conventional analysis' have been developed within the framework of linear theory, namely power spectral analysis for 'noise' and eigenvalue analysis for 'flutter'. However, neither is very useful for nonlinear theory and a single procedure is advantageous. For some applications this will be true for linear theory as well.

5.2 Generalized random forces

Before discussing the equations of motion per se we consider the only essentially new feature of the noise problem over the flutter problem, the evaluation of the generalized random forces. If the random pressures are known from measurement then it is a straightforward matter to compute this quantity. Even if only a statistical description of the random pressures, e.g., power spectra or correlations, is given, pressure time histories may be simulated with the same statistical properties. Shinozuka [68] has given a particularly efficient method for generating generalized random forces which has been further improved by Vaicaitis, *et al.* [69]. This is the present state-of-the-art. Here we would like to digress briefly to consider the possibility of obtaining the required information from a theoretical analysis of the basic fluid mechanics.

A turbulent flow is produced by a laminar flow becoming unstable, i.e., the laminar flow 'flutters'. Of course this is a purely aerodynamic or fluid mechanical instability. Hence it seems natural to attempt to use mathematical methods to treat the nonlinear limit cycle oscillations of unstable laminar flow, 'turbulent flow', similar to those which have been successfully employed for flutter of plates. The first efforts along these lines are reported in [70–71] where, two-dimensional, incompressible Poiseuille channel flow was studied. This is clearly a physical situation of limited practical interest but one which lends itself to relatively simple analysis. In [70] the procedure was to take the (Navier-Stokes) equation of fluid motion in terms of a stream function, ψ, and expand ψ in terms of a series of modes satisfying the fluid boundary conditions at the walls of the channel. Using Galerkin's method the result is a system of nonlinear ordinary differential equations for the time-dependent modal coefficients which may be solved by numerical integration. Hellums, *et al.* [71], have used a finite difference method, rather than a modal approach, and make a strong case that this is the method of choice for this problem in terms of computational efficiency. While much remains to be done to include three-dimensionality, compressibility and extend the calculations to higher Reynolds number, the initial results are encouraging. The basic difficulty is the large number of modes or difference elements which participate in the limit cycle which constitutes turbulent fluid motion.

5.3 Solution procedure and results

Returning to the problem of determining structural response, consider for definiteness Equations (3.2-10a and b); these equations suitably non-dimensionalized are a system of ordinary nonlinear integral-differential equations for the q_n as function of time, t. The solution procedure is to specify the various non-dimensional problem parameters (see Chapter 6 for further discussion), e.g., λ^*, μ, M, a/b, ν, ζ_d, P, λ_c^*, μ_c, d/a, P_r, etc. and to determine the q_n/h by numerical integration of Equation (3.2-10a and b) and, subsequently, w/h from Equation (3.2-9). Stresses in the plate, cavity pressures and other quantities of interest may also be determined.

It is convenient to define certain spatial-temporal averages which describe the gross response of the system. We will define

$$\overline{\left(\frac{w}{h}\right)^2} \equiv \lim_{T\to\infty} \frac{1}{T} \int_0^T \int_0^1 \int_0^1 (w/h)^2 \frac{\mathrm{d}x}{a} \frac{\mathrm{d}y}{b} \mathrm{d}t$$

and

$$\overline{\left(\frac{p_c}{\rho_c a_c^2}\right)^2} \equiv \lim_{T\to\infty} \frac{1}{T} \int_0^T \int_0^1 \int_0^1 \int_{-1}^0 \left(\frac{p_c}{\rho_c a_c^2}\right)^2 \frac{\mathrm{d}x}{a} \frac{\mathrm{d}y}{b} \frac{\mathrm{d}z}{d} \mathrm{d}t$$

These spatial-temporal mean squares are useful measures for comparing the overall response of a plate or cavity under varying parameter conditions.

Extensive numerical results will not be presented here. Instead only a few results will be given to illustrate the method. We are in the fortunate position of being able to assess the importance of the various parameters by using the extensive results already available for the flutter problem. From these the following qualitative conclusion may be reached:

1. λ^* will be important if it is a significant fraction of the flutter dynamic pressure, λ_F^*.

2. Both structural and aerodynamic damping contribute to the total damping of the system. The total damping is always important when the response is that of a resonant structure to 'noise', however, it may be useful to assess which component of the damping is the more important. The usual modal structural damping term is (expressed in a non-dimensional form)

$$2\zeta_n K_n \frac{dq_n}{d\tau}$$

No such simple representation of the aerodynamic damping is generally possible. However for high Mach number one may use 'piston theory', i.e., the integral terms of Equation (2.3) may be neglected and the second term of that expression identified as the aerodynamic damping which gives an aerodynamic damping term (non-dimensionalized) whose magnitude is of the order

$$\frac{(\lambda^* \mu)^{\frac{1}{2}}}{M} \frac{dq_n}{d\tau}$$

Comparing the structural and aerodynamic terms we conclude that μ will be important if

$$\frac{(\mu\lambda^*)^{\frac{1}{2}}}{M} \gtrsim 2\zeta_n K_n$$

i.e., aerodynamic damping is comparable to the structural damping. A converse conclusion applies to ζ_n. This estimate is based upon a high Mach number approximation, say $M > 2$. For $M < 2$ it will generally be necessary to obtain numerical results for any given case.

3. M will be important if λ^* and/or μ are important.

4. a/b is important for $a/b \geqq 1$.

5. $R_x^{(E)}$ and $R_y^{(E)}$, in-plane loads, are important if either is a significant fraction (in absolute magnitude) of the Euler buckling load for the plate.

6. P is important for $P \gtrsim K_1^2$, i.e., if the plate deflection under the static pressure load is on the order of the thickness, $w/h \simeq 1$.

7. $\lambda_c^* a/d$ is important when of the order of K_1^2, the plate (non-dimensional) fundamental frequency squared, i.e., the Helmholtz resonator acoustic spring stiffness is comparable to the plate bending stiffness.
8. μ_c is not generally important.
9. d/a, see above comment on λ_c^*.
10. P_r is always important unless the plate is fluttering, in which case the estimate on P_r is roughly the same as on P.

The conclusions should be modified concerning λ_c^*, μ_c, and d/a by indicating that if the cavity is deep these parameters may be individually significant for determining the cavity pressure field but not for the plate motion.

Finally overall conclusions may be drawn as follows:
1. If λ^* or μ or M is important then the aerodynamic pressure change due to plate motion (aeroelastic feedback) is important.
2. If P is important, or $R_x^{(E)}$, $R_y^{(E)}$ exceed the plate buckling load, or P_r is comparable to P when it is important, or the plate is fluttering, then plate nonlinearities are important.
3. If $\lambda_c^* a/d$ is important then the cavity pressure change due to plate motion (acousto-structural feedback) is important.

If a parameter is not important in the above sense then it may be set to zero without great loss of accuracy.

Turning now to a few illustrative numerical examples, we will first consider the following dimensional values:

$$a = 10 \text{ in.}, \quad b = 20 \text{ in.}, \quad h = 0.03 \text{ in.}, \quad d = 5 \text{ in.},$$

$$\text{Aluminum plate } \rho_m = 0.1 \text{ lb/in}^3, \quad E = 10^7 \text{ lb/in}^2, \quad \beta_n = 0,$$

$$M = 2, \quad \text{altitude } 50{,}000 \text{ ft}, \quad q = 4.73 \text{ lb/in}^2, \quad a_\infty = 971 \text{ ft/sec},$$

$$U_\infty = 1942 \text{ ft/sec}.$$

The random pressure load mean square value is taken from experiment as convected 'frozen white noise' with

$$(\overline{p_r^2})^{\frac{1}{2}} = .006q$$

see, e.g., [27]. The other two parameters which characterize the 'frozen white noise' pressure loading are the convection speed, U_c, and the spatial (or time) interval over which the pressure field is assumed constant, Δx. For these quantities we take $U_c = U_\infty$ and $\Delta x = 2$ in. A uniform probability density is assumed. For a more realistic random load and results therefrom, see Vaicaitis [69].

The above dimensional quantities lead to the following non-dimensional parameters:

$$\lambda^* = 383;$$

$$\mu = 0.0224;$$

$$M = 2;$$

$$a/b = 0.5, \quad d/a = 0.5;$$

$$\zeta_n = 0;$$

$$R_x^{(ES)} = R_y^{(ES)} = P = 0;$$

$$\lambda_c^* = 96;$$

$$\mu_c = 0.0224;$$

$$(\overline{P_r^2})^{\frac{1}{2}} = 380, \quad U_c/U_\infty = 1, \quad \frac{\Delta x}{a} = 0.2.$$

For these calculations the Helmholtz resonator cavity theory is used, in view of the shallow cavity, and piston theory aerodynamics in view of the high Mach number. It might also be mentioned that a reasonable amount of structural damping, $\zeta_n < 0.02$, has no effect on the present results since the aerodynamic or radiation damping dominates.

The mean square values of plate displacement and cavity pressure *vs.* time are given in Fig. 5.2. As may be seen the integration must be carried forward in time until the transient has disappeared, if accurate mean square values are to be obtained. The transient decay time is determined by the amount of damping in the system. In Fig. 5.3 a portion of the time history of the plate displacement is given. The plate response is a random vibration with an average period between the periods of the first two natural modes, 0.51 and 0.15. *If there were no boundary layer 'noise' input the response would be identically zero.*

Now let us consider a change in altitude from 50,000 ft to 30,000 ft. The following values now apply (with all other quantities remaining the same):

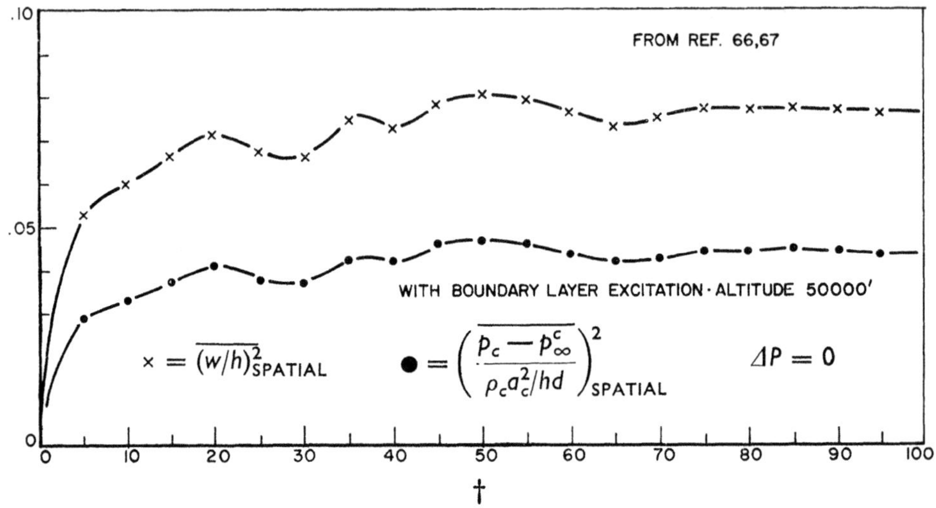

Fig. 5.2 Mean square response vs. time.

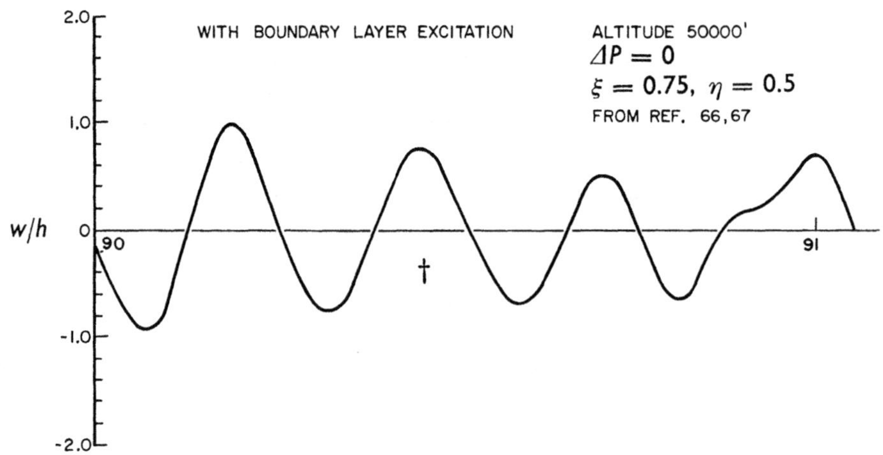

Fig. 5.3 Time history of plate response.

altitude 30,000 ft, $q = 12.2$ lb/in^2,

$a_\infty = 995$ ft/sec, $U_\infty = 1990$ ft/sec,

$\rho_\infty = 0.000858$ slug/ft^3, $U_c = 1990$ ft/sec,

$\rho_c = 0.000858$ slug/ft^3

and in non-dimensional terms:

$\lambda^* = 960;$

$\mu = 0.055;$

$\lambda_c^* = 246;$

$\mu_c = 0.055;$

$$\overline{(P_r^2)}^{\frac{1}{2}} = 960; \quad \frac{U_c}{U_\infty} = 1$$

In Fig. 5.4 the plate response without the boundary layer 'noise' is shown. This is 'flutter', i.e. the linear system is unstable and the result is a non-linear, limit cycle oscillation. $\lambda^*_{\text{flutter}}$ for this plate is roughly 660. As may be seen, the response is essentially a sine wave. Now consider the case with the boundary layer 'noise' present (Fig. 5.5). Here the response of the plate is a combination of self-excited flutter limit cycle oscillation and forced externally excited oscillation due to the boundary layer pressure fluctuations.

Finally, consider the effect of a static pressure differential across the plate. Takes $p^{FS} = 0.1$ lb/in^2 or $P^{FS} = 1350$ which is a representative value. We first present in Fig. 5.6 the response for no boundary layer 'noise'. The pressure differential is large enough so that the flutter motion is completely suppressed and there is simply a static deflection under the static pressure load. In Fig. 5.7 boundary layer 'noise' is present and there is a random oscillation of the plate about the static deflection shape.

It is clear that a pressure differential can have an important effect in suppressing or reducing the plate response. Note however (cf. Fig. 5.5 and 5.7) that though the response amplitude may be reduced the frequency may increase. Hence the effect may not be entirely beneficial as regard fatigue life, for example.

For obvious reasons the examples chosen are ones where:

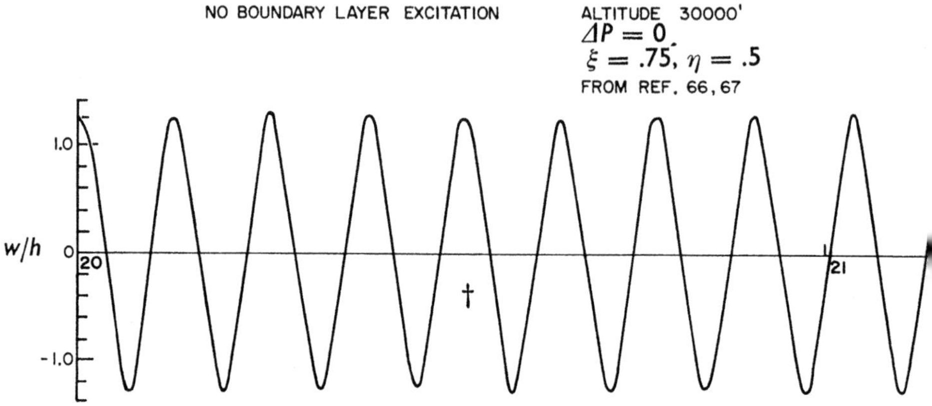

Fig. 5.4 Time history of plate response.

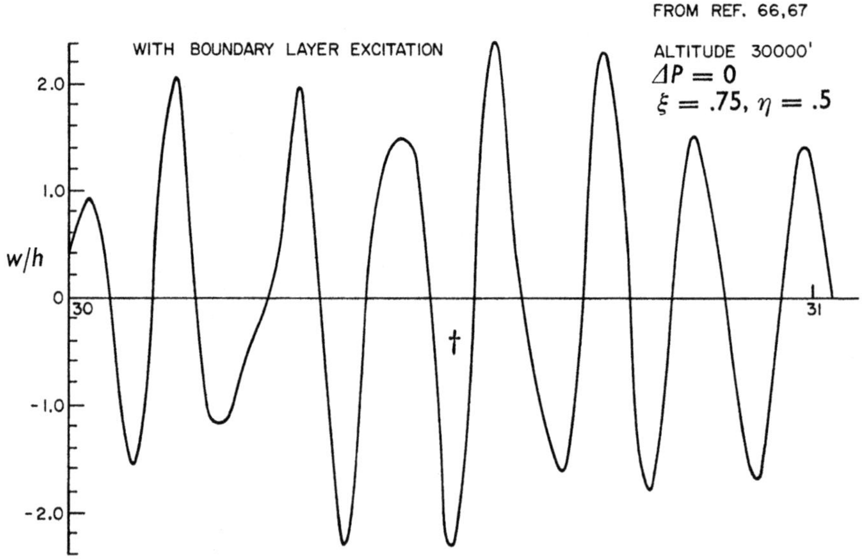

Fig. 5.5 Time history of plate response.

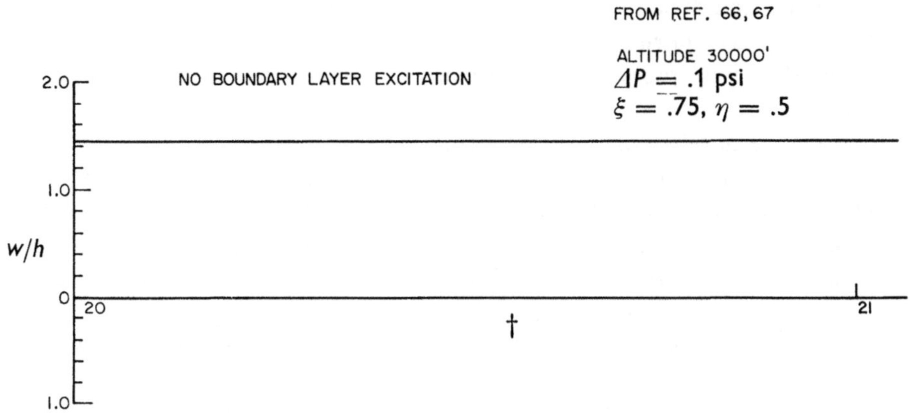

Fig. 5.6 Time history of plate response.

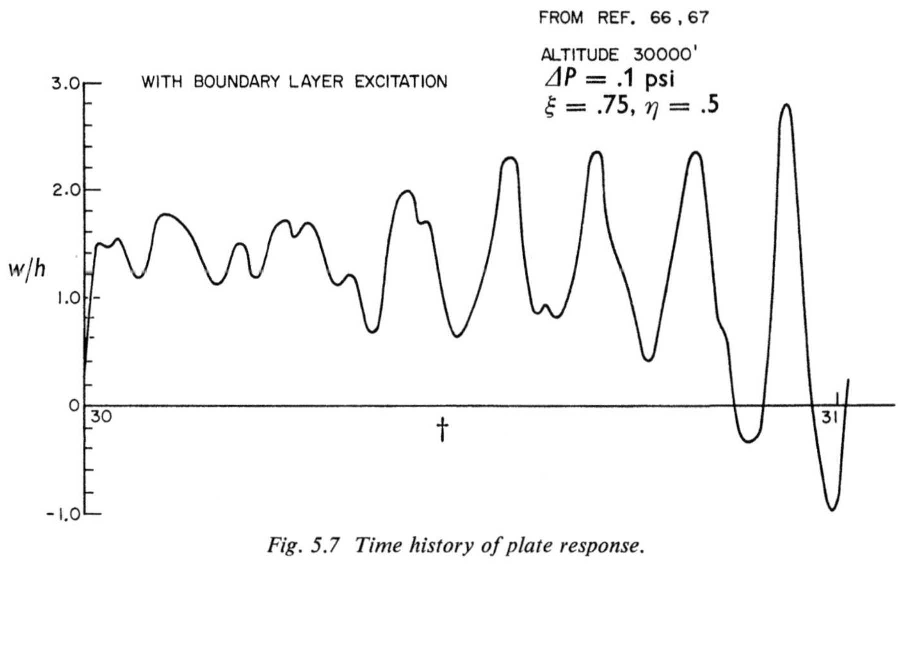

Fig. 5.7 Time history of plate response.

(a) the aerodynamic pressure change due to plate motion, 'aeroelastic feedback', is important;

(b) the cavity pressure on the plate, 'acousto-structural feedback', is important;

(c) plate nonlinearities are important.

Omitting any of these would have led to gross errors for the examples chosen. While it is clearly possible to pick equally reasonable examples where not all or even none of these effects are important, in many practical situations at least one of these effects is apt to be important. The principal improvements needed in the above analysis are

(1) nothing less than a rational, accurate theory of turbulent fluid motion or, in lieu of (1),

(2) systematic exploitation of simulation techniques for turbulent pressure fluctuations as discussed in [68] and [69].

6

Correlation of theory and experiment

In discussing representative theoretical-experimental results, we have two purposes. Firstly, this provides an opportunity to demonstrate to the reader the variety of results which are available in the referenced literature and, secondly, it permits an assessment of the accuracy one may expect to obtain with presently available theoretical and experimental methods.

6.1 Non-dimensional similarity parameters and scaling

In the several theoretical developments, we have carried out the analysis in dimensional terms. While this is preferable to a non-dimensional treatment for ease of understanding, non-dimensionalization does have the advantages of expressing the various results in a most compact form and also establishing scaling laws to extrapolate or interpolate available results to other physical situations.

An appropriate set of non-dimensional parameters may be established by dimensional analysis or more simply by non-dimensionalization of the given structural and aerodynamic equations.

Let us first consider the aerodynamic theory. By examination of the equations we see that an appropriate set of non-dimensional variables would be

$$\begin{aligned} \frac{\Delta p^M}{\rho_\infty U_\infty^2} &= \frac{\Delta p^M}{\rho_\infty U_\infty^2}\left[\frac{x}{a}, \frac{y}{b}, \frac{tU_\infty}{a}; M, a/b, \delta/a, w/a\right] \\ \frac{Q_n^M}{\rho_\infty U_\infty^2 ab} &= \frac{Q_n^M}{\rho_\infty U_\infty^2 ab}\left[\frac{tU_\infty}{a}; M, a/b, \delta/a, w/a\right] \end{aligned} \tag{1}$$

For simplicity we have considered a rectangular plate of length, a, and width, b. This can be generalized to arbitrary shapes. The dependence on the three parameters, M, a/b and δ/a, is explicit while the dependence on the variables x/a, y/b, tU/a, is both explicit and also implicit through the structural motion, w/a. Note that a particular mean velocity profile shape and ratio of specific heats, γ, are also implied.

Similarly we may write for the cavity acoustics

$$\frac{p_c}{\rho_c a_c^2} = \frac{p_c}{\rho_c a_c^2}\left[\frac{x}{a}, \frac{y}{b}, z/a, \frac{ta_c}{a}; a/b, a/d, w/a\right]$$
$$\frac{Q_n^c}{\rho_c a_c^2 ab} = \frac{Q_n^c}{\rho_c a_c^2 ab}\left[\frac{ta_c}{a}; a/b, a/d, w/a\right] \tag{2}$$

From the structural theory of a rectangular, isotropic plate

$$\frac{w}{h} = \frac{w}{h}\left(\frac{x}{a}, \frac{y}{a}, \left(\frac{D}{ma^4}\right)^{\frac{1}{2}} t; a/b, \nu, \frac{\Delta p^M a^4}{Dh},\right.$$
$$\left.\frac{p_c a^4}{Dh}, \frac{\Delta p^{ES} a^4}{Dh}, \frac{p^R a^4}{Dh}\right) \tag{3}$$

and a similar relation could be written for a non-dimensional stress.

Note that w/h also implicitly depends upon the parameters, M, δ/a through $\frac{\Delta p^M}{\rho_\infty U_\infty^2}$ and a/d through $\frac{p_c}{\rho_c a_c^2}$. Moreover by choosing a common non-dimensionalization for pressure, structural deflection and time in (1), (2) and (3) we discover certain other parameters.

$$\frac{\Delta p^M a^4}{Dh} = \frac{\Delta p^M}{\rho_\infty U_\infty^2} \lambda^* \frac{a}{h}$$
$$\frac{p_c a^4}{Dh} = \frac{p_c}{\rho_c a_c^2} \lambda_c^* \frac{a}{h} \tag{4}$$
$$\frac{w}{h} = \frac{w}{a}\frac{a}{h}$$

$$\left(\frac{D}{ma^4}\right)^{\frac{1}{2}} t = (\mu/\lambda^*)^{\frac{1}{2}} \frac{tU_\infty}{a}$$

$$= (\mu_c/\lambda_c^*)^{\frac{1}{2}} \frac{ta_c}{a} \tag{4}$$

where

$$\lambda^* \equiv \frac{\rho_\infty U_\infty^2 a^3}{D} \qquad \mu \equiv \frac{\rho_\infty a}{m}$$

$$\lambda_c^* \equiv \frac{\rho_c a_c^2 a^3}{D} \qquad \mu_c \equiv \frac{\rho_c a}{m}$$

Hence four additional parameters appear, λ^*, μ, λ_c, μ_c. It might seem that a fifth parameter is also present, a/h. However if we use a linear aerodynamic theory such that $\Delta p^M \sim w$, $p^c \sim w$, the factor a/h will cancel out, see (1) and (2). If however we employed a nonlinear version of the aerodynamic theory, then a/h will appear explicitly. As we shall see in a moment a/h will also appear for structural reasons if the plate is curved. Finally we note that while p^R is not at present obtainable from any theory, dimensional analysis considerations suggest that it can be expressed as:

$$\frac{p^R}{\rho_\infty U_\infty^2} = \frac{p^R}{\rho_\infty U_\infty^2}\left[\frac{x}{\delta}, \frac{y}{\delta}, \frac{tU_\infty}{\delta}; M\right] \tag{5}$$

Formally this introduces no new parameters unless one neglects boundary layer effects in Δp^M in which case the above reintroduces δ/a. δ/a itself, of course, would more fundamentally depend upon Reynolds number and Prandtl number as well as wall temperature conditions. See Dugundji and Calligeros [24].

Combining all of the above we may express the explicit dependence of the panel deflection as

$$\frac{w}{h} = \frac{w}{h}\left(\frac{x}{a}, \frac{y}{b}, \left(\frac{D}{ma^4}\right)^{\frac{1}{2}} t;\ \lambda^*, \mu, M, a/b, \delta/a,\right.$$

$$\left.\frac{\Delta p^{ES}}{Dh}, \nu, \lambda_c^*, \mu_c, a/d\right) \tag{6}$$

Recall this is for a rectangular isotropic flat plate with a rectangular cavity. The precise functional form is determined by panel edge support conditions and cavity wall conditions. For other panel and cavity shapes a similar relation would hold with a, b, and d characteristic lengths associated with panel length (chord), width (span) and cavity depth, respectively.

Other parameters may appear if other physical effects are included. Some of these are listed below.

Flow angle of yaw:

$$\theta$$

Externally applied in-plane loads:

$$\frac{N_x^E a^2}{D}, \quad \frac{N_y^E a^2}{D}, \quad \frac{N_{xy}^E a^2}{D}$$

Or for the particular case of thermal in-plane loads

$$12(1 + \nu)\left(\frac{a}{h}\right)^2 \alpha \Delta T$$

Curvature:

$$\frac{z_p}{h}(x, y), \; a/h$$

where z_p is the initially curved shape of the plate. For a cylindrical shell a/h might be replaced by R/h where R is the radius of the shell and a/b would be replaced by a/R.

In conclusion we should make several points. First of all the particular choice of non-dimensional parameters offered here is not unique but it has proven useful. The reader will note we have employed these parameters in the presentation of our various results. Secondly, we have not given non-dimensional parameters for some physically interesting effects, e.g., orthotropicity, support flexibility and structural damping, although such can be constructed. Finally, the more general but still relevant discussion of [24] is again recommended to the reader.

6.2 Flutter results for particular structural geometries and loading conditions

Flat Plate: With the exception of [18] no experimental data have been obtained with the express purpose of investigating the post-flutter (nonlinear) regime for any type of panel. In determining the flutter boundary, however, some limited data in this regime have been taken. A plot of peak deflection amplitude versus dynamic pressure for a flat plate [12] is shown in Fig. 6.1. Note the plate thickness was .041 inches. Below the flutter dynamic pressure q_F, the plate amplitude is very small compared to the plate thickness while for $q > q_F$ the amplitude becomes comparable to the thickness. This is in qualitative agreement with theory but additional data to larger dynamic pressure well beyond the flutter boundary are needed for an adequate quantitative comparison. Also see [18].

With data of the type shown in Fig. 6.2 one may determine (within a certain error) the flutter dynamic pressure. Available quantitative comparisons between theory and experiment are primarily for q_F. In Fig. 6.2 such comparisons are shown for plates of three different thicknesses. In the figure legend the three numbers are length in inches, width in inches and thickness in thousandths of an inch, respectively. In dimensional form, we have q_F

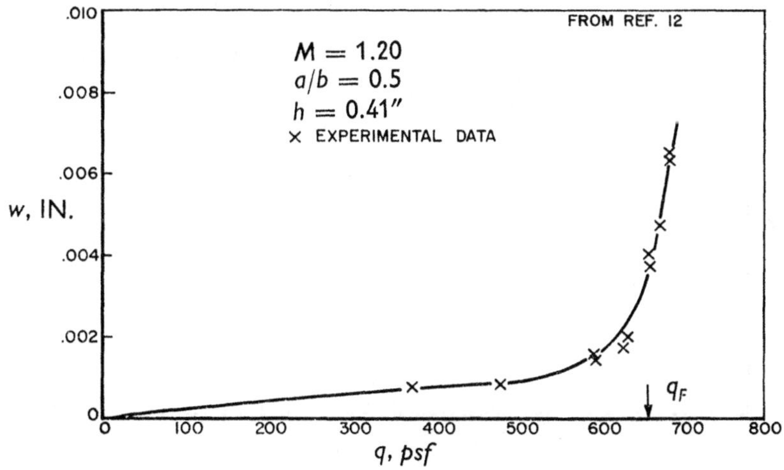

Fig. 6.1 Peak panel response vs. dynamic pressure.

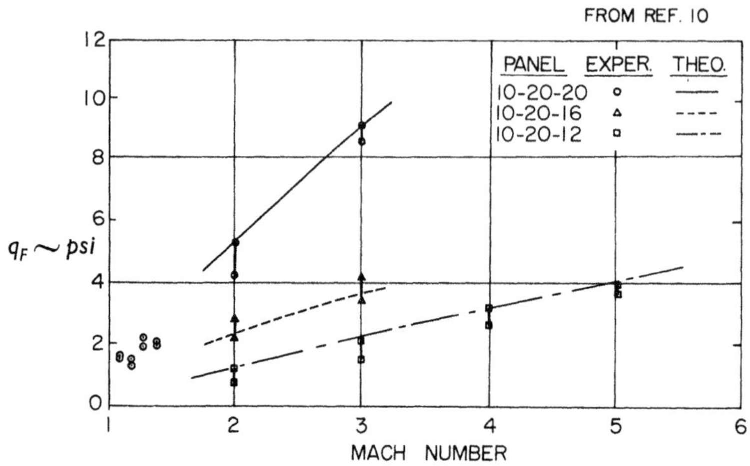

Fig. 6.2 Flutter dynamic pressure vs. Mach number.

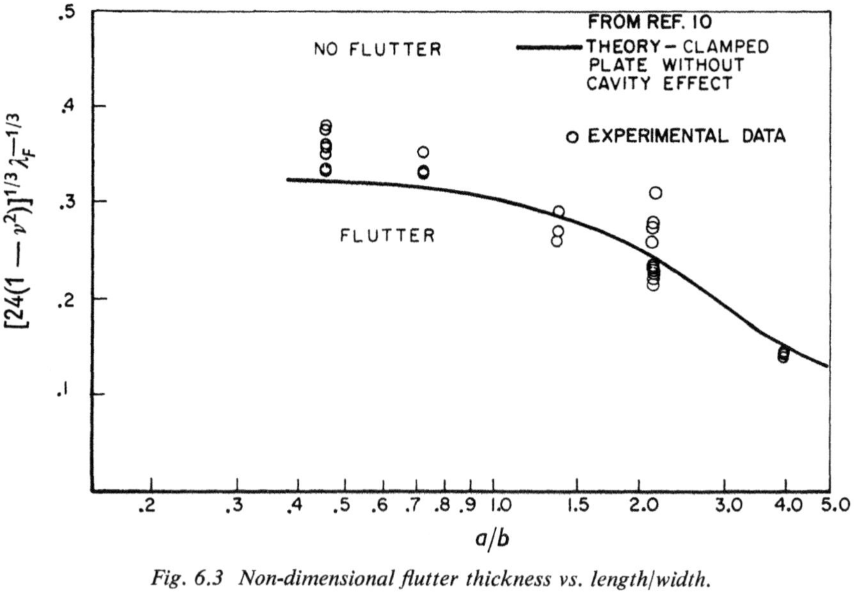

Fig. 6.3 Non-dimensional flutter thickness vs. length/width.

versus M. As may be seen the agreement between theory and experiment is quite satisfactory. These data are for $a/b = 0.46$; similar data with similar agreement are presented in [10] for $a/b = 0.73$, 1.37, 2.18 and 4.11. The plates are aluminum and clamped on all edges. The theory is a linear analysis using 'piston theory' aerodynamics with cavity effect and measured natural frequencies in place of their theoretical counterparts. Most of these data, as in Fig. 6.2, are for $M \geqq 2$. For these high Mach numbers, one may use the correlation parameter λ. In Fig. 6.3 all the experimental data of [10] for $M \geqq 2$ are presented in nondimensional form and compared with theory for an ideal clamped plate, neglecting the cavity. The parameter actually shown in the figure $\lambda^{-\frac{1}{3}}$ is essentially the thickness/length ratio required to prevent flutter (above the boundary). This parameter, often used in the literature, is more commonly employed than λ itself for design situations. Each experimental point represents a different plate and Mach number. As may be seen agreement between theory and experiment is generally satisfactory, particularly in view of the large number of dimensional variables contained in λ. The scatter shown in the experimental data is due to a number of factors but primarily is a reflection of construction differences in the several test models and the neglect of the cavity effect in the theory. Construction differences could result in a $\pm 10\%$ scatter and the cavity in $+5 \rightarrow 10\%$ difference of theory from experiment. Additional data of this type are available which are also in reasonable agreement with theory; see Dixon [72] for an interesting and useful discussion of theoretical experimental correlation.

A smaller amount of data in [10] for $M < 2$ shows poorer agreement with theory. Recently, Muhlstein and Gaspers, *et al.* [12, 13], have conducted experiments to determine flutter boundaries of thin, flat plates at low supersonic Mach number, $M = 1.1$–1.4. Their careful, systematic examination of aerodynamic boundary layer effects on panel flutter has shown this to be the primary source of the differences between theory and experiment found in [10].

In Fig. 6.4 the experimental data of [12] and the theory of [55] for a length-to-width ratio of one-half, $a/b = 0.5$, are presented in terms of flutter dynamic pressure, λ_F^*, versus Mach number, M. Experimental data for zero boundary layer thickness (as extrapolated) and also the maximum boundary layer thickness of the tests are shown with corresponding theoretical curves. The plate edges are clamped. As can be seen, theory and experiment are in generally good agreement. Additional comparisons are needed to establish greater confidence in the theory.

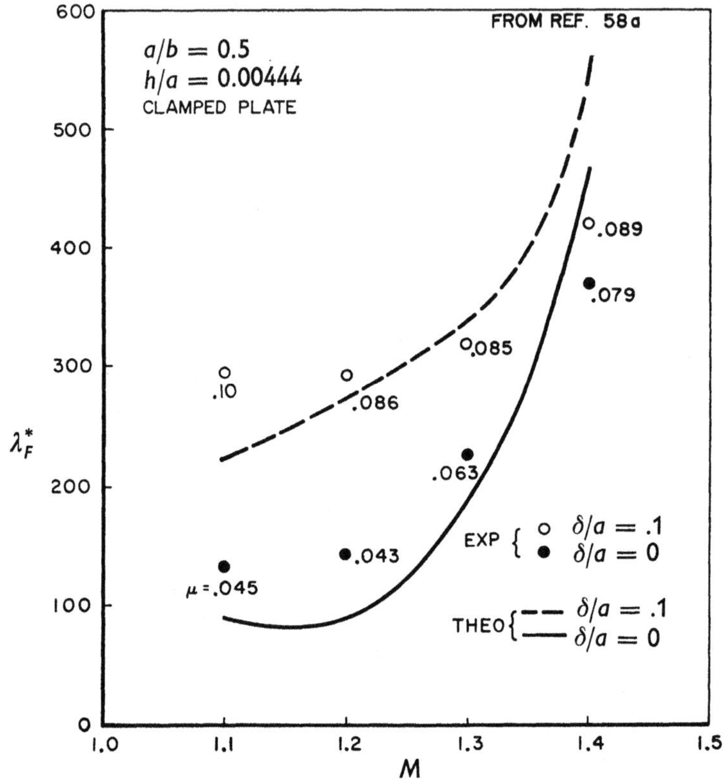

Fig. 6.4 Flutter dynamic pressure vs. Mach number.

For $a/b = 10$ limited available theoretical, linear flutter analyses, and experimental data for $M = 3$, also show reasonable agreement. There are indications from theory that, for large a/b, q_F becomes independent of panel length and a new parameter $\lambda_b^* = 2qb^3/D$ with a replaced by b would be appropriate. This also seems physically plausible. Generally speaking, plates of large a/b are relatively unexplored. Again see [72].

Flat plate under transverse and in-plane loads

Pressure Loaded Plates: A plate exposed to a transverse static pressure load undergoes a deformation that causes stretching in the middle surface of the plate. The membrane stresses associated with this stretching effectively

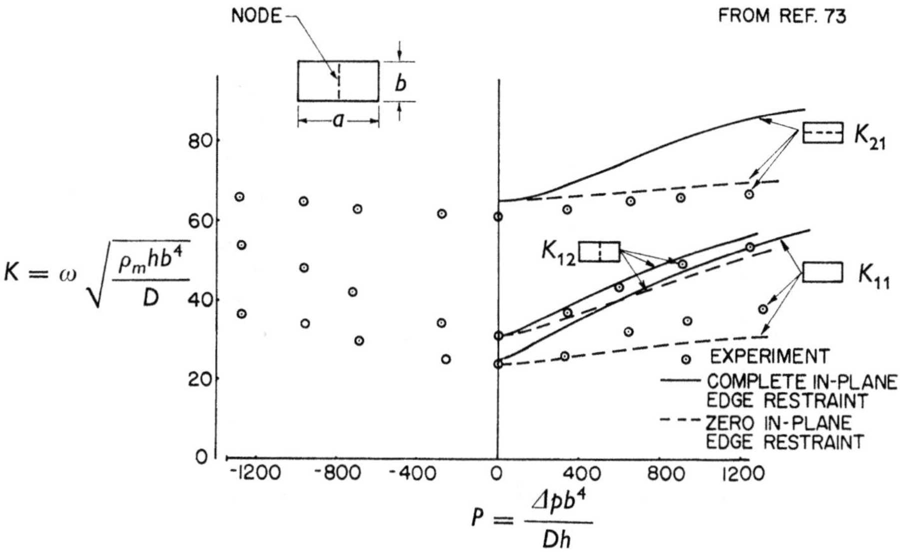

Fig. 6.5 Natural frequencies vs. static pressure differential.

stiffen the plate, raising its natural frequencies of vibration. In-plane boundary support flexibility will relieve the stresses and thereby reduce the increase in frequencies of the loaded plate. The effect of a static pressure differential on the natural frequencies and flutter boundary of a panel has been studied in some detail by Ventres [73]. Shown in Fig. 6.5 are the natural frequencies computed from a Galerkin analysis for the limiting cases of zero and complete in-plane restraint and also those determined experimentally. Note that theory and experiment agree quite well at $\Delta p = 0$, but that the assumption of complete in-plane constraint considerably overestimates the frequencies for $\Delta p \neq 0$. The experimental data lie closest, in fact, to the curves for zero in-plane restraint. Presumably then, the supports to which the plates edges were bonded were sufficiently rigid to restrain them against rotation (thereby maintaining a clamped boundary condition), yet were unable to provide complete restraint against in-plane motions at the edges. Comparison of calculated and measured static deflections further confirmed this.

Flutter boundaries have been determined as a function of the pressure differential between the cavity and a static pressure tap in front of the panels. The measured flutter boundaries were not symmetric about $\Delta p = 0$, pre-

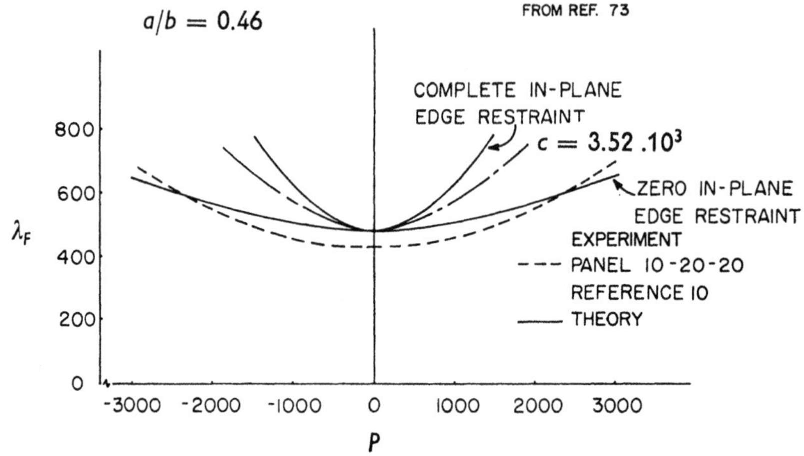

Fig. 6.6 Flutter dynamic pressure vs. static pressure differential.

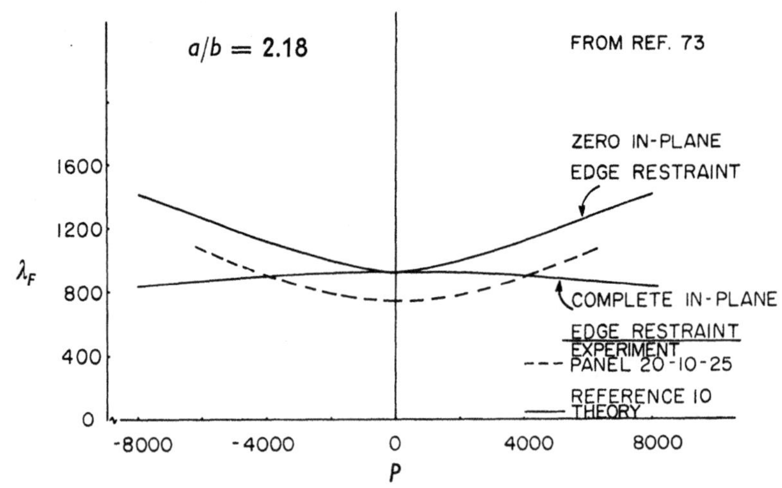

Fig. 6.7 Flutter dynamic pressure vs. static pressure differential.

sumably due to pressure gradients over the plates or to flow misalignment [10]. Therefore, the data presented (in dimensionaless form) in Figs. 6.6 and 6.7 have been adjusted to make them symmetric about $\Delta p = 0$. This effectively assumes the existence of some equivalent constant Δp.

In calculating theoretical flutter boundaries for comparison with these results, the effect of the cavity has been included. In addition, experimentally determined values for the natural frequencies (for $\Delta p = 0$) of the two panels, taken from [10], were used to account approximately for the effects of imperfect panel construction. These frequencies characterize the linear elastic behavior of the panels. They were inserted into the modal equations in place of their theoretical counterparts to correct the linear stiffness of the plates.

Note that the relative positions of the flutter boundaries for zero and for complete in-plane restraint are reversed in the two figures. Some understanding of these rather surprising results can be obtained by studying the natural frequency spectra. Natural frequencies of a plate under a pressure load depend not only on the membrane stresses caused by the load, but on the induced middle surface curvature as well. For stiffly restrained plates, the effect of the tension is predominant, but for relatively unrestrained plates the influence of the curvature must be considered also. The best agreement between theory and experiment both for natural frequencies and flutter is usually obtained by assuming zero edge restraint which is probably typical for panels of practical construction.

Buckled Plates: In-plane loads can cause a plate to buckle, thereby altering its behavior in a uniform airflow. The buckled plate has a curved middle surface and mid-plane stresses that depend in a complicated manner on the nature of the in-plane loading, the in-plane boundary support conditions, and the external flow. The flutter motion is no longer simple harmonic and may not even be periodic. The buckled plate is frequently a critical design condition for plates subjected to large in-plane compressive loads.

Most experimental work has been done on plates buckled by uniform thermal expansion. [23] is one of the most recent in a series of experiments by NASA Langley. Comparisons have been made between a nonlinear structural theory using 'piston theory' aerodynamics and experimental results from [23] for plates with of $a/b = 2.9$ buckled by thermal expansion [73]. The testing was done at a Mach number of 3, and at constant dynamic pressure. The plates were heated by the airflow so that the in-plane loading increased with time, exceeding the buckling value before termination of the run. With but few exceptions flutter occurred before buckling, and then ceased at some point after buckling.

In this sort of test the in-plane boundary conditions affect not only the large-deflection behavior of the plate, but the nature of the in-plane loading as well. If the edges are completely restrained against in-plane motion, the in-plane stress resultants N_x^E and N_y^E are equal and

$$N_x^E a^2/D = N_y^E a^2/D = 12(1 + \nu)(a/h)^2 \alpha \Delta T$$

where ΔT is the uniform temperature differential between the plate and its support. The non-dimensional temperature parameter

$$R_T \equiv 12(1 + \nu)(a/h)^2 \alpha \Delta T$$

is used to describe the thermal loading on the plate, even though it may not be fully restrained. If the plate is not fully restrained, then the in-plane stresses due to thermal expansion will generally be unequal. The plates tested in [23] were riveted to their support frames.

The assumption of complete in-plane edge restraint produces theoretical flutter boundaries in very poor quantitative agreement with experiment. Improved agreement has been obtained by selecting values of in-plane edge restraint so that buckling would occur for the same R_T experimentally and theoretically. This is equivalent to assuming that the airflow does not affect the static Euler buckling load and allows for the incomplete restraint of the test panels. Using nonlinear stiffness terms based on such assumptions, the stability boundaries shown in Fig. 6.8 were computed [73].

In Fig. 6.8 two flutter boundaries are shown in the buckled regime, i.e., beyond R_T for which λ_f is a minimum. The 'small disturbance flutter boundary' was computed using a buckled equilibrium configuration as an initial condition for the numerical integration of the full nonlinear modal equation. On the other hand, calculations proceeding from a flat plate initial condition produced sustained flutter-like oscillations below this linear flutter boundary. Since the buckled shapes involved deflections on the order of several plate thicknesses, this is physically equivalent to applying a 'large disturbance' to the plate. The numerical results obtained suggest that sustained flutter oscillations are possible for λ, R_T combinations above the shaded area in Fig. 6.8. Note that this 'large disturbance' flutter boundary is in excellent agreement with the experimental data, since most of the data

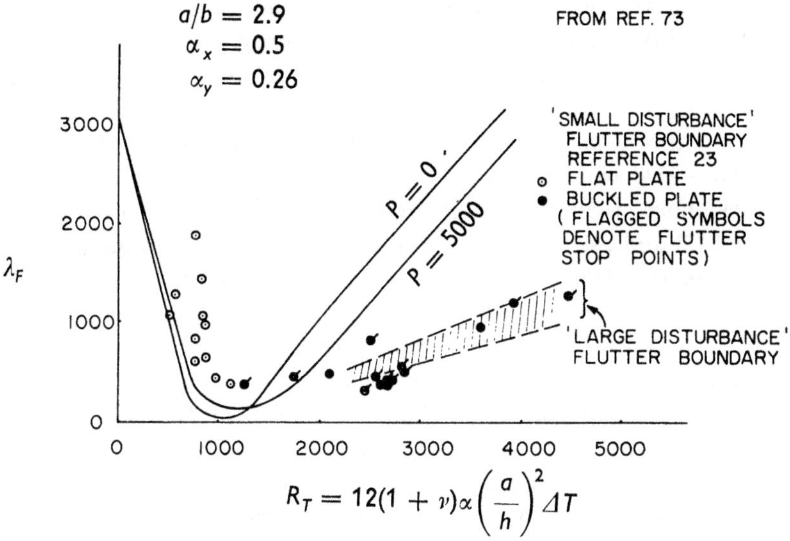

Fig. 6.8 Flutter dynamic pressure vs. temperature differential.

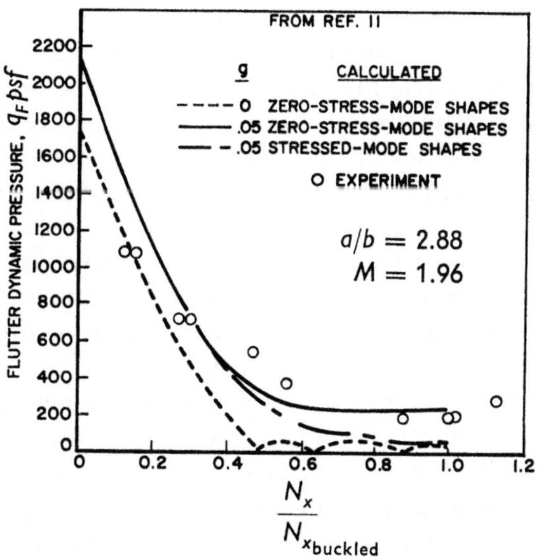

Fig. 6.9 Flutter dynamic pressure vs. in-plane load.

in that region are flutter stop points. On the other hand, the 'small disturbance' boundary is unconservative. The frequency of flutter (not shown) is in generally good agreement with the experimental data for all values of R_T.

The most recent and extensive work on the flutter of buckled plates is that of Hess [11] who conducted an experimental and analytical study of flat panel under streamwise mechanical compressive loads for a range of $a/b = 1$ to 4.2 and $M = 1.57$ and 1.96. He also briefly examined the effects of pressurization. Typical of his results are those given in Fig. 6.9 where flutter dynamic pressure is plotted versus $\sigma/\sigma_c = N_x^{ES}/N_{xc}^{ES}$, the applied in-plane stress non-dimensionalized by the critical Euler buckling stress. Both experimental data and the results from theory employing three different assumptions concerning structural damping are shown. In the 'zero stressed mode shapes' case the structural damping, $1 + ig$, multiplied only the linear bending terms while in the 'stressed mode shapes' theory all stiffness terms were so modified. All analyses used linear plate theory and 'piston theory' aerodynamics. As can be seen, the theoretical results are sensitive to damping near minimum q_f. The results are probably also sensitive to panel imperfections [74] but the requisite calculations have not been carried out for this panel. Hess suggests the use of structural damping to obtain realistic results for buckled plates based on these results.

Shore [42, 75] has also conducted theoretical and experimental studies on the effect of structurally damping on thermally buckled plates and has suggested design charts [76] for flutter of panels on the verge of buckling based upon linear plate theory, 'piston theory' aerodynamics and an 'appropriate' model of structural damping, essentially that for 'zero stressed mode shapes'. Whatever the rigorous standing of structural damping as the reconciler of theory and experiment, such charts are useful within the inherent limitations of the structural and aerodynamic theories employed.

Most recently Ventres [77] has used the nonlinear flutter theory to compute theoretical results to compare with the Hess data for a pressurized *and* buckled plate. The agreement is surprisingly good.

Curved Plate: The experimental data which are available, due to Anderson [78], are for square plates with streamwise curvature clamped on four edges. The available theoretical results are for two-dimensional curved clamped plates (including the static aerodynamic loading) [20]. In Fig. 6.10 the experimental and theoretical flutter boundaries are shown in terms of λ versus H/h where H is maximum rise of curved plate above

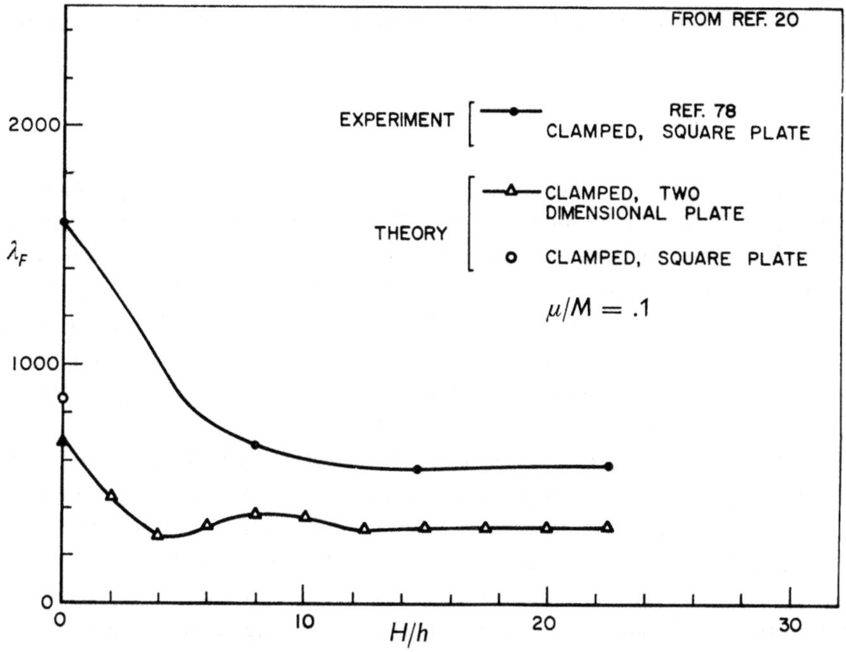

Fig. 6.10 Flutter dynamic pressure vs. plate rise for a streamwise curved plate.

supports and

$$z_p = 4H \frac{x}{a} \left[1 - \frac{x}{a} \right]$$

There is qualitative agreement between the results of theory and experiment as can be seen. Interestingly enough the largest discrepancy is for a flat plate. Also shown, is a theoretical result for a flat square plate. Note this only gives a modest improvement in the agreement between theory and experiment for $H/h = 0$. The discrepancies are attributed largely to experimental uncertainties in this early test (1962).

Two other features of the flutter motion determined theoretically are in good qualitative agreement with the experimental flutter data. The theoretical flutter amplitude is on the order of H in good agreement with experiment. Also the theoretical plate deformation shapes under static aerodynamic load just below the flutter boundary are pulled out toward the trailing edge and pressed in toward the leading edge, again agreeing with the experimental evidence [20]. Generally streamwise curvature leads to lower flutter dynamic pressures.

Cylindrical shell:

Stearman *et al.* [79], and later Olson [80] have conducted experiments at high supersonic Mach number, $M = 3$. Stearman [81] has also reported additional tests at low supersonic speeds, $M = 1.2$–1.5. The high Mach number experimental data have been compared to the theoretical results of [80] and [50] which use piston aerodynamic theory and a linear structural model. See Fig. 6.11 taken from [50] which gives λ_f versus a static pressure parameter. There is general agreement between theory and experiment on the character of the flutter oscillation, a high circumferential mode number flutter, see [50] and [80]. The agreement on flutter dynamic pressure is also reasonably good for an unpressurized shell but becomes poor as the static pressure differential is increased. Caution is also advised in view of the use of piston theory fluid forces which are much less accurate for cylindrical shell than a flat plate [57, 82]. Olson [80] has also used the full linearized potential aerodynamic forces on an infinitely long cylinder and a finite

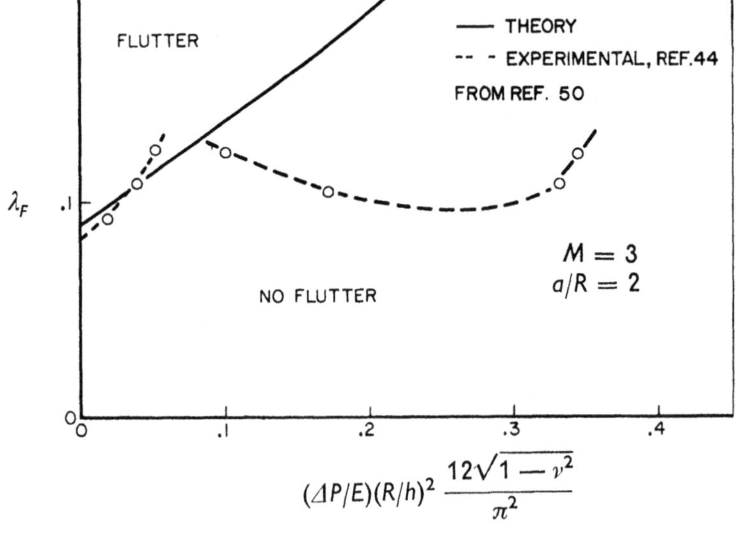

Fig. 6.11 Flutter dynamic pressure vs. static pressure loading for a cylindrical shell.

length structural shell analysis which leads to much poorer agreement with experiment.

For the low supersonic regime, the experimental data indicated the flutter mode is of a low circumferential mode number, traveling wave type. Hence, Stearman [81] has compared the results with those of the infinitely long shell analysis of Miles [83] which was developed for this type of flutter. Generally, the agreement is not good for flutter dynamic pressure, the theoretical values being much less then the experimental ones. The theory of [84] gives a similar result. The slender body approximation, which has been suggested for high circumferential mode number flutter [57], was also used by Stearman to compare with this experiment and, as might be expected, is in poor agreement with the experimental data giving a value of flutter dynamic pressure considerably higher than the experimental data.

Despite the fact that the fluid forces on a finite length cylinder have been determined using the full linearized potential theory [57], only various asymptotic approximations to these forces have been used even in the linear shell flutter analyses until the recent work of LeClerc [46] and Johns [47]. Their results are valuable but cover a limited parameter range. One may anticipate that the more complete aerodynamic theory will be used to a greater extent in the future for both linear and nonlinear analyses. Hopefully, this will help clarify the rather complicated picture which presently exists with respect to the relation of theory and experiment for cylindrical shells. Matsuzaki and Kobayashi [82] have developed an aerodynamic theory intermediate in complexity and accuracy between the 'exact' and 'slender body' theories of [57]. No flutter calculations have been reported as yet but it may well be that this represents a useful compromise between the two extremes.

With regard to improvements in the structural model, Barr and Stearman [51] have shown that cylinders are sensitive to structural imperfections on the order of the shell thickness. Although the precise quantitative nature of the imperfections are unknown, qualitatively speaking they appear to account for some of the differences between theory and experiment.

Also Matsuzaki and Kobayashi [85] have reported an experimental and theoretical study of the unpressurized cylindrical shell at $M = 2$. Their results indicate that linear shell theory with quasi-steady, two-dimensional aerodynamics is satisfactory for predicting the onset of flutter but that the nonlinear shell theory is required to accurately predict the flutter mode and frequency.

It is generally found that cross-stream curvature leads to higher flutter dynamic pressure than that for a flat plate.

Finally, the important work of Paidoussis should be mentioned. In the most recent and comprehensive of a series of papers [86] he has studied theoretically and experimentally the flutter of a cylindrical shell with an *internal, low speed* fluid flow. Such configurations are of interest in nuclear reactor technology. This work cannot be directly related to the studies discussed above which were concerned with flutter in an external, high speed fluid flow. However, it is of great interest in the present context because of generally good correlation obtained between theory and experiment as well as for the insight it gives into the interesting change in type of instability (divergence or flutter) and mode of instability (tubular beam mode or shell mode) with shell length and boundary conditions. In this respect, see Appendix II for further discussion.

Other physical parameters

Orthotropicity: Considerable experimental and theoretical work has been done by NASA Langley on orthotropic panels. The most recent report of this work is contained in [87]. Representative of their results is Fig. 6.12. Non-dimensional dynamic pressure is plotted versus K_d or $1/K_d$ where K_d

$$\lambda_F = \frac{2q_F b^3}{D_1\sqrt{M^2-1}}\left(\frac{D_1}{D_{12}}\right)^{3/2}$$

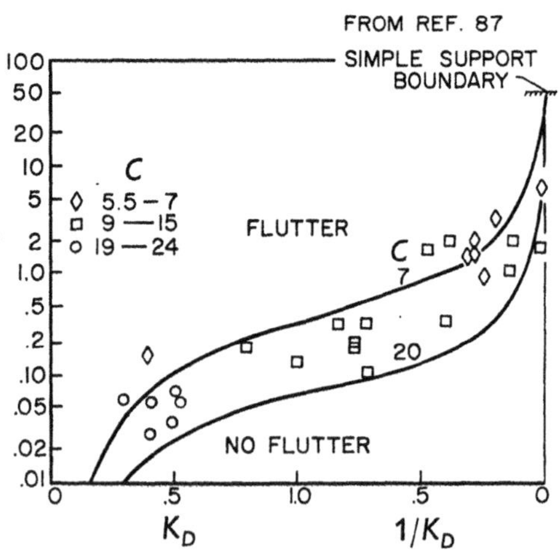

Fig. 6.12 Flutter dynamic pressure vs. support stiffness.

is a non-dimensional support stiffness along the streamwise edges. D_1, D_2 are orthotropic stiffnesses in the streamwise and spanwise directions and D_{12} the twisting stiffness. $1/K_d \to 0$ represents a simply-supported edge and $K_d \to 0$ a free edge. C is the ratio of twisting to (the square root of the product of) bending stiffness; see [87] for details. $D_1/D_2 \ll 1$ and two solid theoretical curves are shown for $C = 7$ and 20. Plotted points are experimental data. The important point is that many practical orthotropic panel configurations have significant support flexibility along edges perpendicular to the direction of largest stiffness (along the streamwise edge in the results of Fig. 6.12). Careful determination of boundary conditions (support stiffness) becomes very important for such panels since the flexible edge may lead to flutter dynamic pressures much below those for a rigid, simply-supported edge.

A valuable case history for such panels is that of the X-20A (hypersonic research aircraft) project. See [88] for a discussion of experimental-theoretical calculations. In this instance it was found necessary to use measured panel natural frequencies in flutter analyses in order to obtain good agreement with flutter tests. This example also points up the difficulty of measuring natural modes and frequencies of complicated built-up panels with sufficient accuracy and in sufficient number adequate for flutter analysis or correlation with flutter test. In the vibration tests, impedance techniques were used wherein a natural mode resonance was inferred by the (near) vanishing of the required exciting force.

Angle-of-yaw: Experimental data are reported in [89] including comparisons with theory for several types of orthotropic plate. For the simpler panels correlation is reasonable but for the more complicated panels agreement is less satisfactory probably due to the difficulties associated with orthotropicity rather than angle-of-yaw per se. On physical grounds, the angle-of-flow effect should be adequately described by theory if an appropriate aerodynamic analysis is employed.

Omitted Parameters: No extensive, experimental data exist for edge conditions (except in the context of orthotropic plates, see previous discussion) planform geometry, cavity effect, structural damping, geometrical imperfections, or multibay panels.

It would appear that edge conditions and planform geometry effects can be adequately described theoretically if their effect on panel natural modes and frequencies can be predicted. The cavity effect is thought to be adequately modeled by theory on the basis of prediction of panel/cavity natural frequencies [10]. If measured values of structural damping and/or geometric imperfections are used, then theory is probably reliable at least for qualitative trend studies. Multibay effects are not likely to be important but the theory [90–92] for multibay effects should be fundamentally no more nor less reliable than for a single isolated panel.

Traveling wave flutter

The controversy in the early literature surrounding standing wave versus traveling wave flutter was in part a matter of semantics but also involved substance as well. Any standing wave can be thought of as a combination of traveling waves and vice versa, of course. Moreover one would never find a pure standing wave or traveling wave in any experimental situation.

A standing wave is a wave form with fixed node lines wherein each point on the plate undergoes simple harmonic motion and there is no change in spatial phase with time. Mathematically

$$w = \mathrm{Re}\, \bar{w}(x) \mathrm{e}^{\mathrm{i}\omega t} \qquad \mathrm{Re} \equiv \text{Real Part}$$

A traveling wave is a wave form with node lines which move with constant velocity. Mathematically

$$w = w(x - ct)$$

where c is the (constant) wave speed.

Most flutter motions are wave forms with moving node lines which move with a changing wave speed.

The substantial question is, which of the two ideal wave forms most closely approaches those observed in practice? It turns out that for most flutter motions the movement of the node lines is small compared to a typical plate dimension, say plate length or width, and hence one has essentially a standing wave. However there are at least two known cases where the motion has a very clear traveling wave character. The first is that of an incompressible flow over a plate on an elastic foundation which has been thoroughly studied by Dugundji [93], *et al.* Their work shows good agreement between theory and experiment with respect to flutter dynamic pressure and wavelength with somewhat poorer agreement on frequency and wavespeed. The flutter is of a ‘merging wave speed’ type with flutter essentially occurring when the backward and forward wave speeds present at zero dynamic pressure are driven together by increasing q. Until recently this particular case was considered to be an interesting but somewhat academic curiosity whose principal virtue was to clearly establish that traveling wave flutter was possible. However Srinivasan [94] and colleagues [95] have clearly established another case of traveling wave flutter wherein a cylinder in a circumferential, rotating vortex-like flow has a flutter motion which takes the form of a traveling wave around the circumference of the cylinder. This case has applications in high speed turbine technology. Because of the complex nature of the mean flow, the development of an accurate aerodynamic and concomitant flutter theory is somewhat involved. However theory and experiment have shown reasonable agreement.

6.3 Post-flutter behavior and structural response to turbulent boundary layer pressure fluctuations

Post-flutter behavior

Ventres and Kang [96] have used a non-linear structural model (Von Karman's plate equation) with potential flow aerodynamic theory to calculate the limit cycle flutter oscillations of the plates studied experimentally by Kappus, Lemley and Zimmerman [18]. The post-flutter, limit cycle data from [18] are for a large length/width plate ($a/b \cong 4.5$) in a nearly buckled condition. The agreement between theory and experiment is only qualitative with better agreement being obtained for maximum values of stress amplitudes. That is, theory and experiment give somewhat different values for the point or location on the panel where the maximum stress occur. Hence, if maximum stresses are compared without regard to their particular location the agreement between theory and experiment is improved. Nevertheless, the agreement cannot be considered satisfactory at present and it would be very helpful to make similar comparisons for smaller length/width plates in an unbuckled condition.

Structural response to turbulent boundary layer pressure fluctuations

The most recent results available (which incorporate the lessons learned by earlier investigators, e.g. Maestrello [97] and Wilby [98]) are due to Chyu and Au-yang [99, 100]. For rectangular, flat, clamped panels of moderate length/width ratio they have measured the random structural response. Comparisons have been made with theoretical structural response results which employ as input measured boundary layer pressure fluctuations, measured structural damping and calculated aerodynamic forces due to structural motion. The latter provide 'aeroelastic feedback' damping to the plate. Comparisons between theory and experiment have also been made for this important quantity for subsonic Mach numbers in [100] (using a potential aerodynamic model) and supersonic Mach numbers in [58b] using (a shear flow aerodynamic model and) test data from Muhlstein. Generally, agreement between theory and experiment for 'aeroelastic damping' is satisfactory except in the transonic flow regime. 'Aeroelastic damping' typically dominates structural damping at high subsonic Mach numbers and above. The agreement between theory and experiment for structural response per se is satisfactory when the damping is accurately known.

7

Concluding remarks

In this monograph we have sought to provide a connected account of the aeroelasticity of plates and shells. Several major themes have been explored beyond those normally encountered in the standard aeroelastic literature: the importance of structural non-linearities, the necessity for accounting for shear (boundary layer) effects in the fluid model under certain circumstances and the relationship of forced externally excited vibration (due to turbulent boundary layer pressure fluctuations, 'noise') to self-excited vibration (due to aeroelastic instability, 'flutter'). In these concluding remarks several additional themes are introduced which have not yet been adequately explored, here or elsewhere.

It will be clear to the reader that some of the techniques discussed herein could be usefully employed in other areas of aeroelasticity, notably the aeroelastic behavior of lifting surfaces as affected by nonlinearities and viscous flow effects. However, the requisite applications are not simple and will require significant advances in our present capabilities. Nevertheless the results described herein should serve as helpful guides to future developments. Not the least of the benefits to be derived is the revival of the notion that stability and response questions might best be treated together rather than separately.

The determination of the turbulent pressure fluctuations themselves from first principles may also benefit from the ideas explored here. While we have taken these as given, in point of fact they are the limit cycle ('flutter', if you will) oscillations of an unstable (laminar) flow. To adequately model this instability some of the ideas explored herein will perhaps prove useful. Preliminary efforts have already been made in this direction. What is needed, of course, is nothing less than a rational theory of turbulent fluid motion which has eluded us for almost a century.

Finally the close interplay between experimental and theoretical techniques for the aeroelasticity of plates and shells has attained a standard of effectiveness which hopefully will have an impact on various aeroelastic problems beyond those considered here.

Appendix I

Elastic strain energy and various forms of the equilibrium equations

Here we shall review some of the basic relations for the behavior of elastic, shallow shells. This discussion roughly follows that of Volmir [101, but also see Bolotin [2] and Washizu [102]. For a linear elastic isotropic material, the stress-strain relations are:

$$\begin{aligned}
\sigma_x &= \frac{E}{(1-\nu^2)}[\varepsilon_x + \nu\varepsilon_y] \\
\sigma_y &= \frac{E}{(1-\nu^2]}[\varepsilon_y + \nu\varepsilon_x] \\
\sigma_{xy} &= G\varepsilon_{xy}
\end{aligned} \tag{1}$$

The strai n-displacement relations for a shallow curved plate are:

$$\begin{aligned}
\varepsilon_x &= \frac{\partial u}{\partial x} + \frac{1}{2}\left(\frac{\partial w}{\partial x}\right)^2 + \frac{\partial z_p}{\partial x}\frac{\partial w}{\partial x} - z\frac{\partial^2 w}{\partial x^2} \\
\varepsilon_y &= \frac{\partial v}{\partial y} + \frac{1}{2}\left(\frac{\partial w}{\partial y}\right)^2 + \frac{\partial z_p}{\partial y}\frac{\partial w}{\partial y} - z\frac{\partial^2 w}{\partial y^2} \\
\varepsilon_{xy} &= \frac{\partial v}{\partial x} + \frac{\partial u}{\partial y} + \frac{\partial w}{\partial x}\frac{\partial w}{\partial y} + \frac{\partial z_p}{\partial x}\frac{\partial w}{\partial y} + \frac{\partial z_p}{\partial y}\frac{\partial w}{\partial x} - 2z\frac{\partial^2 w}{\partial x \partial y}
\end{aligned} \tag{2}$$

For a linear elastic solid the elastic energy is given by

$$U = \tfrac{1}{2}\iiint [\sigma_x\varepsilon_x + \sigma_y\varepsilon_y + \sigma_{xy}\varepsilon_{xy}]\mathrm{d}_x\mathrm{d}_y\mathrm{d}z \tag{3}$$

Define

$$
\begin{aligned}
N_x &\equiv \int \sigma_x \mathrm{d}z \\
N_y &\equiv \int \sigma_y \mathrm{d}z \\
N_{xy} &\equiv \int \sigma_{xy} \mathrm{d}z \\
\varepsilon_{xo} &\equiv \frac{\partial u}{\partial x} + \frac{1}{2}\left(\frac{\partial w}{\partial x}\right)^2 + \frac{\partial z_p}{\partial x}\,\frac{\partial w}{\partial x} \\
\varepsilon_{yo} &\equiv \frac{\partial v}{\partial y} + \frac{1}{2}\left(\frac{\partial w}{\partial y}\right)^2 + \frac{\partial z_p}{\partial y}\,\frac{\partial w}{\partial y} \\
\varepsilon_{xyo} &\equiv \frac{\partial v}{\partial x} + \frac{\partial u}{\partial y} + \frac{\partial w}{\partial x}\,\frac{\partial w}{\partial y} + \frac{\partial z_p}{\partial x}\,\frac{\partial w}{\partial y} + \frac{\partial z_p}{\partial y}\,\frac{\partial w}{\partial x}
\end{aligned}
\tag{4}
$$

Hence

$$
\begin{aligned}
N_x &= \frac{Eh}{(1-\nu^2)}\left[\varepsilon_{xo} + \nu\varepsilon_{yo}\right] \\
N_y &= \frac{Eh}{(1-\nu^2)}\left[\varepsilon_{yo} + \nu\varepsilon_{xo}\right] \\
N_{xy} &= Gh\varepsilon_{xyo}
\end{aligned}
\tag{5}
$$

and from (1)–(5)

$$
\begin{aligned}
U = {} & \frac{Eh}{2(1-\nu^2)}\iint\left[\varepsilon_{xo}^2 + \varepsilon_{yo}^2 + 2\nu\varepsilon_{xo}\varepsilon_{yo} + \frac{(1-\nu)}{2}\varepsilon_{xyo}^2\right]\mathrm{d}x\,\mathrm{d}y \\
& + \frac{D}{2}\iint\left[\left(\frac{\partial^2 w}{\partial x^2}\right)^2 + \left(\frac{\partial^2 w}{\partial y^2}\right)^2 + 2\nu\,\frac{\partial^2 w}{\partial x^2}\,\frac{\partial^2 w}{\partial y^2}\right. \\
& \left. + 2(1-\nu)\left(\left(\frac{\partial^2 w}{\partial x \partial y}\right)^2\right)\right]\mathrm{d}x\,\mathrm{d}y
\end{aligned}
\tag{6}
$$

The first term of (6) is usually referred to as the stretching energy, U_s, and the second as the bending energy, U_B. Solving for ε_{xo}, ε_{yo}, ε_{xyo}, in terms of N_x, N_y, N_{xy} from (5) the stretching energy may be written as

$$U_s = \frac{1}{2Eh}\iint [N_x^2 + N_y^2 - 2\nu N_x N_y + 2(1+\nu)N_{xy}^2]\mathrm{d}x\mathrm{d}y$$

$$= \tfrac{1}{2}\iint [N_x\varepsilon_{xo} + N_y\varepsilon_{yo} + N_{xy}\varepsilon_{xyo}]\mathrm{d}x\mathrm{d}y \tag{7}$$

Introducing the Airy stress function, F, it may be written as

$$U_s = \frac{1}{2Eh}\iint\Bigg[\left(\frac{\partial^2 F}{\partial x^2}\right)^2 + \left(\frac{\partial^2 F}{\partial y^2}\right)^2 - 2\nu\frac{\partial^2 F}{\partial x^2}\frac{\partial^2 F}{\partial y^2} + 2(1+\nu)\left(\frac{\partial^2 F}{\partial x \partial y}\right)^2\Bigg]\mathrm{d}x\mathrm{d}y \tag{8}$$

Finally U_s may be expressed in terms of displacements u, v, w, by using (4) and (5) in (6) or (7).

$$\begin{aligned} U_s = \frac{Eh}{2(1-\nu^2)}\iint\Bigg\{&\left[\frac{\partial u}{\partial x} + \frac{1}{2}\left(\frac{\partial w}{\partial x}\right)^2 + \frac{\partial z_p}{\partial x}\frac{\partial w}{\partial x}\right]^2 \\ &+ \left[\frac{\partial v}{\partial y} + \frac{1}{2}\left(\frac{\partial w}{\partial y}\right)^2 + \frac{\partial z_p}{\partial y}\frac{\partial w}{\partial y}\right]^2 \\ &+ 2\nu\left[\frac{\partial u}{\partial x} + \frac{1}{2}\left(\frac{\partial w}{\partial x}\right)^2 + \frac{\partial z_p}{\partial x}\frac{\partial w}{\partial x}\right] \left[\frac{\partial v}{\partial y} + \frac{1}{2}\left(\frac{\partial w}{\partial y}\right)^2 + \frac{\partial z_p}{\partial y}\frac{\partial w}{\partial y}\right] \\ &+ \frac{(1-\nu)}{2}\Bigg[\frac{\partial v}{\partial x} + \frac{\partial u}{\partial y} + \frac{\partial w}{\partial x}\frac{\partial w}{\partial y} \\ &+ \frac{\partial z_p}{\partial x}\frac{\partial w}{\partial y} + \frac{\partial z_p}{\partial y}\frac{\partial w}{\partial x}\Bigg]^2\Bigg\}\mathrm{d}x\mathrm{d}y \end{aligned} \tag{9}$$

Equation (6) using (9) is probably the most useful form of the strain energy. Equations (3.2-1) and (3.2-4) are derivable from (6) and (7) taking variations with respect to u, v, w and using equations (4) and (5) to identify the resulting terms in (3.2-4). Natural boundary conditions are also obtained. Note equations (3.2-1) and (3.2-2) are not derivable from (6) and (8) using Hamilton's Principle with variations with respect to w and F. This is because F is implicitly a function of w, see (4) and (5). To derive (3.2-2) we replace N_x, N_y, N_{xy} in (5) by appropriate derivatives of F, and ε_{xo}, etc. in (5) using (4). The resulting three equations for F, u, v, may be used to obtain (3.2-2) by appropriate differentiation and elimination of u and v.

We shall outline in some detail the above derivations and at the end briefly discuss the necessary modifications to account for orthotropicity and thermal stresses. Taking the variation of (6) and using (5) gives

$$\begin{aligned}\delta U = &\iint [N_x \delta\varepsilon_{xo} + N_y \delta\varepsilon_{yo} + N_{xy}\delta\varepsilon_{xyo}]\,\mathrm{d}x\,\mathrm{d}y \\ &+ D\iint\left[\frac{\partial^2 w}{\partial x^2}\,\delta\left(\frac{\partial^2 w}{\partial x^2}\right) + \frac{\partial^2 w}{\partial y^2}\,\delta\left(\frac{\partial^2 w}{\partial y^2}\right)\right. \\ &+ \nu\frac{\partial^2 w}{\partial x^2}\,\delta\frac{\partial^2 w}{\partial y^2} + \nu\frac{\partial^2 w}{\partial y^2}\,\delta\frac{\partial^2 w}{\partial x^2} \\ &\left.+ 2(1-\nu)\frac{\partial^2 w}{\partial x \partial y}\,\delta\frac{\partial^2 w}{\partial x \partial y}\right]\mathrm{d}x\,\mathrm{d}y \end{aligned} \tag{10}$$

Now from (2)

$$\begin{aligned}\delta\varepsilon_{xo} &= \delta\left(\frac{\partial u}{\partial x}\right) + \left(\frac{\partial w}{\partial x} + \frac{\partial z_p}{\partial x}\right)\delta\left(\frac{\partial w}{\partial x}\right) \\ \delta\varepsilon_{yo} &= \delta\left(\frac{\partial v}{\partial y}\right) + \left(\frac{\partial w}{\partial y} + \frac{\partial z_p}{\partial y}\right)\delta\left(\frac{\partial w}{\partial y}\right) \\ \delta\varepsilon_{xyo} &= \delta\left(\frac{\partial v}{\partial x}\right) + \delta\left(\frac{\partial u}{\partial y}\right) + \left(\frac{\partial w}{\partial x} + \frac{\partial z_p}{\partial x}\right)\delta\left(\frac{\partial w}{\partial y}\right) \\ &\quad + \left(\frac{\partial w}{\partial y} + \frac{\partial z_p}{\partial y}\right)\delta\left(\frac{\partial w}{\partial x}\right)\end{aligned} \tag{11}$$

Using (11) in (10) and integrating the several terms by parts and collecting them we obtain

$$\delta U = -\iint\left\{\left[\frac{\partial N_x}{\partial x} + \frac{\partial N_{xy}}{\partial y}\right]\delta u\right.$$

$$+\left[\frac{\partial N_y}{\partial y} + \frac{\partial N_{xy}}{\partial x}\right]\delta v$$

$$+\left[\frac{\partial}{\partial x}\left(N_x\left(\frac{\partial w}{\partial y} + \frac{\partial z_p}{\partial x}\right)\right)\right.$$

$$+\frac{\partial}{\partial y}\left(N_y\left(\frac{\partial w}{\partial y} + \frac{\partial z_p}{\partial y}\right)\right)$$

$$+\frac{\partial}{\partial y}\left(N_{xy}\left(\frac{\partial w}{\partial x} + \frac{\partial z_p}{\partial x}\right)\right)$$

$$\left.\left.+\frac{\partial}{\partial x}\left(N_{xy}\left(\frac{\partial w}{\partial y} + \frac{\partial z_p}{\partial y}\right)\right)\right]\delta w\right\}\mathrm{d}x\,\mathrm{d}y$$

$$+\iint\left\{N_x\delta u + N_{xy}\delta v\right.$$

$$\left.+\left[N_x\left(\frac{\partial w}{\partial x} + \frac{\partial z_p}{\partial x}\right) + N_{xy}\left(\frac{\partial w}{\partial y} + \frac{\partial z_p}{\partial y}\right)\right]\delta w\right\}\Big|_0^a \mathrm{d}y$$

$$+\iint\left\{N_y\delta v + N_{xy}\delta u\right.$$

$$\left.+\left[N_y\left(\frac{\partial w}{\partial y} + \frac{\partial z_p}{\partial y}\right) + N_{xy}\left(\frac{\partial w}{\partial x} + \frac{\partial z_p}{\partial x}\right)\right]\delta w\right\}\Big|_0^b \mathrm{d}x$$

$$+ D\iint \nabla^4 w\,\delta w\,\mathrm{d}x\,\mathrm{d}y$$

$$+ D\int\left\{\left[\frac{\partial^2 w}{\partial x^2} + \nu\frac{\partial^2 w}{\partial y^2}\right]\delta\frac{\partial w}{\partial x}\right.$$

$$+\left[-\frac{\partial^3 w}{\partial x^3}-(2-\nu)\frac{\partial^3 w}{\partial x\partial y^2}\right]\delta w\Bigg\}\Bigg|_0^a \mathrm{d}y$$

$$+D\int\Bigg\{\left[\frac{\partial^2 w}{\partial y^2}+\nu\frac{\partial^2 w}{\partial x^2}\right]\delta\frac{\partial w}{\partial y}$$

$$+\left[-\frac{\partial^3 w}{\partial y^3}-(2-\nu)\frac{\partial^3 w}{\partial y\partial x^2}\right]\delta w\Bigg\}\Bigg|_0^b \mathrm{d}x$$

$$+D\,2(1-\nu)\frac{\partial^2 w}{\partial x\partial y}\delta w\Bigg|_0^a\Bigg|_0^b$$

The variation of the kinetic energy and the virtual work done by the several aerodynamic pressures similarly may be evaluated as

$$\delta T=\iint m\frac{\partial w}{\partial t}\delta\left(\frac{\partial w}{\partial t}\right)\mathrm{d}x\mathrm{d}y \tag{13}$$

$$\delta W=-\iint \Delta p\,\delta w\,\mathrm{d}x\mathrm{d}y \tag{14}$$

Combining (12), (13) and (14) into Hamilton's Principle

$$\int[\delta T-\delta U+\delta W]\mathrm{d}t=0 \tag{15}$$

and integrating δT by parts with respect to time we obtain, by setting the coefficient of each independent variation to zero,

$$\delta u:\ \frac{\partial N_x}{\partial x}+\frac{\partial N_{xy}}{\partial y}=0 \tag{16}$$

on $x=0, a$ $\qquad N_x=0$ or u specified

on $y=0, b$ $\qquad N_{xy}=0$ or u specified

$$\delta v:\ \frac{\partial N_y}{\partial y}+\frac{\partial N_{xy}}{\partial x}=0 \tag{17}$$

on $x = 0, a$ $\quad\quad\quad\quad N_{xy} = 0$ or v specified

on $y = 0, b$ $\quad\quad\quad\quad N_y = 0$ or v specified

$$\delta w: D\nabla^4 w - N_x \frac{\partial}{\partial x}\left(\frac{\partial w}{\partial x} + \frac{\partial z_p}{\partial x}\right) - N_y \frac{\partial}{\partial y}\left(\frac{\partial w}{\partial x} + \frac{\partial z_p}{\partial y}\right)$$

$$- N_{xy}\left(\frac{\partial^2 w}{\partial x \partial y} + \frac{\partial^2 z_p}{\partial x \partial y}\right) + m\frac{\partial^2 w}{\partial t^2} + \Delta p = 0 \tag{18}$$

on $x = 0, a$

$$D\left[\frac{\partial^2 w}{\partial x^2} + \nu\frac{\partial^2 w}{\partial y^2}\right] = 0 \quad\quad \text{or } \frac{\partial w}{\partial x} \text{ specified}$$

$$- D\left[\frac{\partial^3 w}{\partial x^3} + (2 - \nu)\frac{\partial^3 w}{\partial x \partial y^2}\right]$$

$$+ N_x\left(\frac{\partial w}{\partial x} + \frac{\partial z_p}{\partial x}\right) + N_{xy}\left(\frac{\partial w}{\partial y} + \frac{\partial z_p}{\partial y}\right) = 0 \quad\quad \text{or } w \text{ specified}$$

on $y = 0, b$

$$D\left[\frac{\partial^2 w}{\partial y^2} + \nu\frac{\partial^2 w}{\partial x^2}\right] = 0 \quad\quad \text{or } \frac{\partial w}{\partial y} \text{ specified}$$

$$- D\left[\frac{\partial^3 w}{\partial y^3} + (2 - \nu)\frac{\partial^3 w}{\partial x \partial y^2}\right]$$

$$+ N_y\left(\frac{\partial w}{\partial y} + \frac{\partial z_p}{\partial y}\right) + N_{xy}\left(\frac{\partial w}{\partial x} + \frac{\partial z_p}{\partial x}\right) = 0 \quad\quad \text{or } w \text{ specified}$$

on $x = 0, a$ and $y = 0, b$

$$D2(1 - \nu)\frac{\partial^2 w}{\partial x \partial y} = 0 \quad\quad \text{or } w \text{ specified}$$

(16) and (17) have been used to simplify (18). In addition for simplicity we have done the analysis for a constant thickness plate and rectangular

boundaries. Thickness variations or more complicated shapes offer no essential difficulty as far as equation derivation is concerned. Compare (18) and (3.2-1).

To establish the strain compatibility equation we form

$$\frac{\partial^2}{\partial y^2}(4d) + \frac{\partial^2}{\partial x^2}(4e) - \frac{\partial^2}{\partial x \partial y}(4f)$$

thus eliminating u and v to obtain

$$\frac{\partial^2 \varepsilon_{xo}}{\partial y^2} + \frac{\partial^2 \varepsilon_{yo}}{\partial x^2} - \frac{\partial^2 \varepsilon_{xyo}}{\partial x \partial y} = - \frac{\partial^2 z_p}{\partial x^2}\frac{\partial^2 w}{\partial y^2} - \frac{\partial^2 z_p}{\partial y^2}\frac{\partial^2 w}{\partial x^2} - \frac{\partial^2 w}{\partial y^2}\frac{\partial^2 w}{\partial x^2} + 2\frac{\partial^2 z_p}{\partial x \partial y}\frac{\partial^2 w}{\partial x \partial y} + \left(\frac{\partial^2 w}{\partial x \partial y}\right)^2 \tag{19}$$

Inverting (1) and using (5) along with definition of the Airy stress-function (note (16) and (17) are identically satisfied by (20)),

$$N_x \equiv \frac{\partial^2 F}{\partial y^2}$$

$$N_y \equiv \frac{\partial^2 F}{\partial x^2} \tag{20}$$

$$N_{xy} \equiv -\frac{\partial^2 F}{\partial x \partial y}$$

we obtain from (19)

$$\frac{\nabla^4 F}{Eh} = \left[\frac{\partial^2}{\partial x \partial y}(w + z_p)\right]^2 - \left(\frac{\partial^2 z_p}{\partial x \partial y}\right)^2 - \frac{\partial^2}{\partial x^2}(w + z_p)\frac{\partial^2}{\partial y^2}(w + z_p) + \frac{\partial^2 z_p}{\partial x^2}\frac{\partial^2 z_p}{\partial y^2} \tag{21}$$

where we have written the right-hand side in an alternate, more compact form. Compare (21) and (3.2-2).

We have now derived all of the equations used at various points in the text excluding the effects of orthotropicity and thermal stresses. Here we shall briefly indicate how these effects may be taken into account.

Thermal Stresses:

(1) becomes

$$\begin{aligned} \sigma_x &= \frac{E}{(1-\nu^2)}[\varepsilon_x + \nu\varepsilon_y - (1+\nu)\alpha T] \\ \sigma_y &= \frac{E}{(1-\nu^2)}[\varepsilon_y + \nu\varepsilon_x - (1+\nu)\alpha T] \end{aligned} \tag{22}$$

(5) becomes

$$\begin{aligned} N_x &= \frac{Eh}{(1-\nu^2)}[\varepsilon_{x_0} + \nu\varepsilon_{y_0}] - \frac{Eh}{(1-\nu)}\alpha T \\ N_y &= \frac{Eh}{(1-\nu^2)}[\varepsilon_{y_0} + \nu\varepsilon_{x_0}] - \frac{Eh}{(1-\nu)}\alpha T \end{aligned} \tag{23}$$

where $T = T(x, y)$ is the temperature distribution over plate assumed constant through the plate thickness.

(21) becomes

$$\frac{\nabla^4 F}{Eh} = -\alpha^2 T + \dots \tag{24}$$

All other equations are unchanged.

Orthotropicity:

(1) becomes

$$\begin{aligned} \sigma_x &= E_x'\varepsilon_x + E''\varepsilon_y \\ \sigma_y &= E_x'\varepsilon_y + E''\varepsilon_x \end{aligned} \tag{25}$$

Omitting intermediate stops, we find that equations (16) and (17) are unchanged and that (18) becomes

$$D_x \frac{\partial^4 w}{\partial x^4} + 2[D_1 + 2D_{xy}] \frac{\partial^4 w}{\partial x^2 \partial y^2} + D_y \frac{\partial^4 w}{\partial y^4} + \ldots = \ldots$$

on $x = 0, a$

$$D_x \frac{\partial^2 w}{\partial x^2} + D_1 \frac{\partial^2 w}{\partial y^2} = 0 \ldots$$

$$- D_x \frac{\partial^2 w}{\partial x^3} - [D_1 + 4D_{xy}] \frac{\partial^3 w}{\partial x \partial y^2} + \ldots = 0$$

on $y = 0, b$

$$D_1 \frac{\partial^2 w}{\partial x^2} + D_y \frac{\partial^2 w}{\partial y^2} = 0 \ldots$$

$$- [D_1 + 4D_{xy}] \frac{\partial^3 w}{\partial y \partial x^2} - D_y \frac{\partial^3 w}{\partial y^2} + \ldots = 0 \ldots$$

on $x = 0, a$ and $y = 0, b$

$$4D_{xy} \frac{\partial^2 w}{\partial x \partial y} = 0 \ldots \tag{26}$$

(21) becomes

$$\frac{1}{h} \left\{ \frac{\left[E_y' \frac{\partial^4 F}{\partial y^4} - 2E'' \frac{\partial^4 F}{\partial x^2 \partial y^2} + E_x' \frac{\partial^4 F}{\partial x^4} \right]}{E_y' E_y' - E''^2} + \frac{1}{G} \frac{\partial^4 F}{\partial x^2 \partial y^2} \right\}$$

$$= \ldots \tag{27}$$

where

$$D_x \equiv \frac{E_x' h^3}{12}$$

$$D_y \equiv \frac{E'_y h^3}{12}$$

$$D_1 \equiv \frac{E'' h^3}{12}$$

$$D_{xy} \equiv \frac{G h^3}{12}$$

Appendix II

Qualitative effect of boundary conditions

In most, though certainly not all, practical applications the structure is restrained at its leading and trailing edges. However there are important distinctions between the case where, for example, (1) both ends are clamped and (2) one end is clamped and the other end free. For simplicity we consider a one-dimensional, flat plate. Further we shall use simplified, idealized aerodynamic pressure relations for convenience, viz.

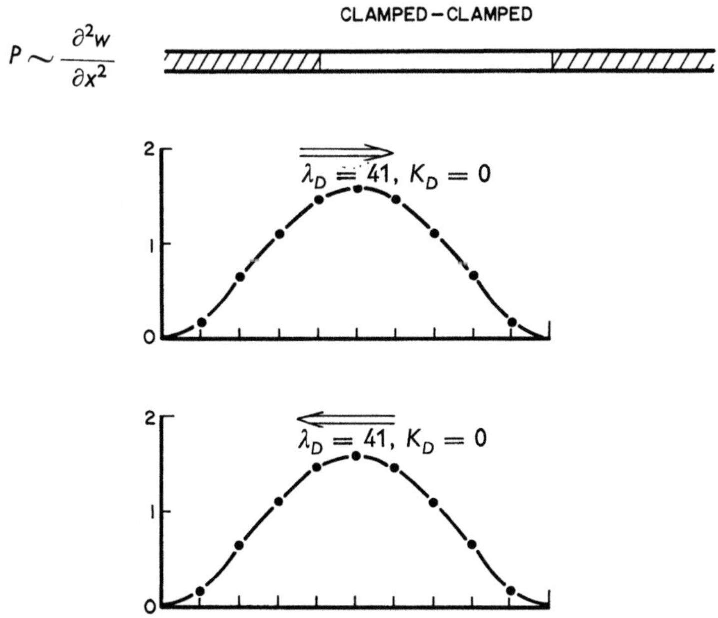

Fig. AII.1

High Mach Numbers:

$$p \sim \frac{\partial w}{\partial x}$$

Low Mach Numbers:

$$p \sim \frac{\partial^2 w}{\partial x^2}$$

This will allow us to consider divergence and merging frequency flutter. Single mode flutter due to negative aerodynamic damping is not qualitatively affected by boundary conditions. Omitting details one finds the results shown in Fig. AII.1 to AII.4, using two-mode Galerkin solutions employing natural modes appropriate to the boundary conditions. As can be seen a change in boundary conditions can lead to a change from flutter to divergence or vice-versa. The results for slender body type aerodynamics are similar to those obtained by Paidoussis [86] and others for internal fluid flows through cylinders.

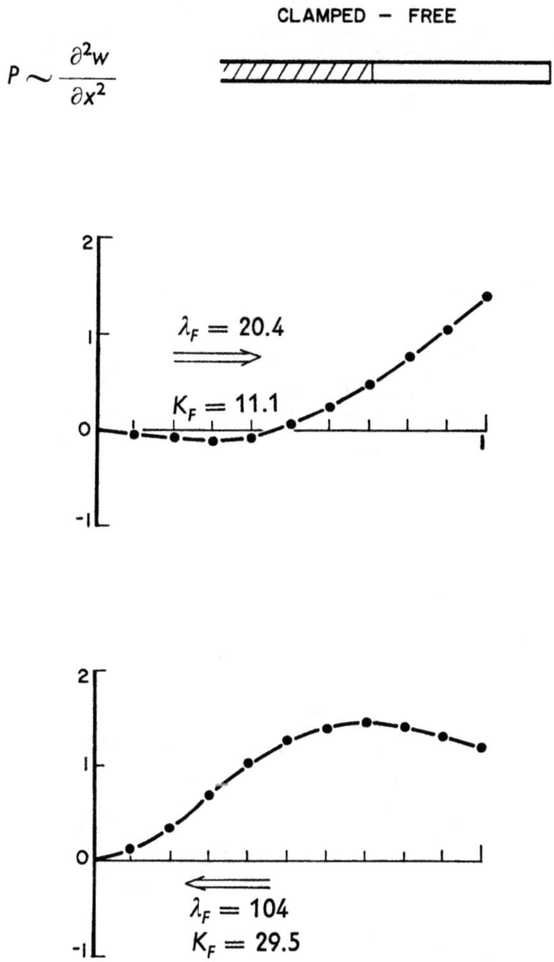
CLAMPED - FREE
$P \sim \frac{\partial^2 w}{\partial x^2}$
2
1
0
-1
$\lambda_F = 20.4$
$K_F = 11.1$
2
1
0
-1
$\lambda_F = 104$
$K_F = 29.5$

Fig. AII.2

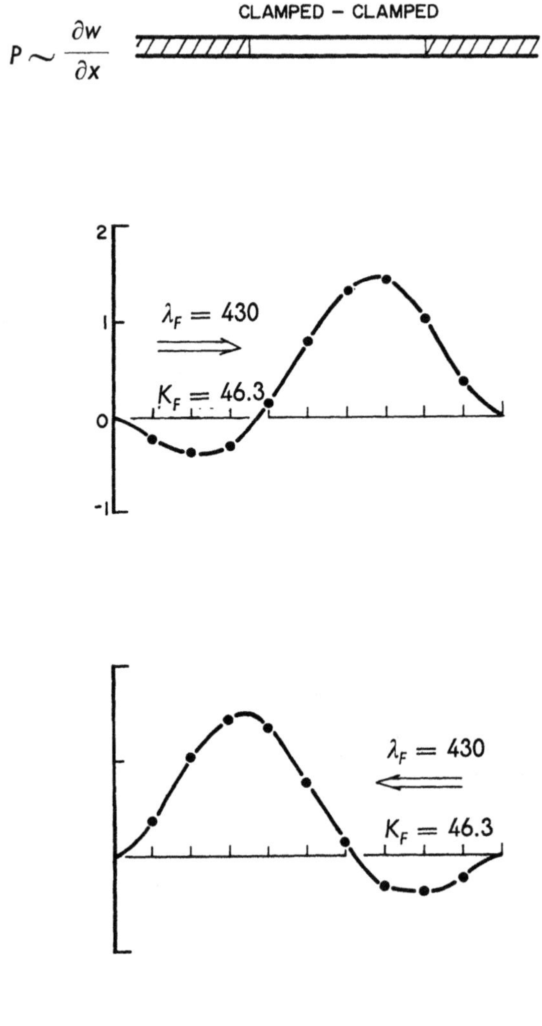
CLAMPED - CLAMPED
P ~ ∂w/∂x
2
1
0
-1
$\lambda_F = 430$
$K_F = 46.3$
$\lambda_F = 430$
$K_F = 46.3$

Fig. AII.3

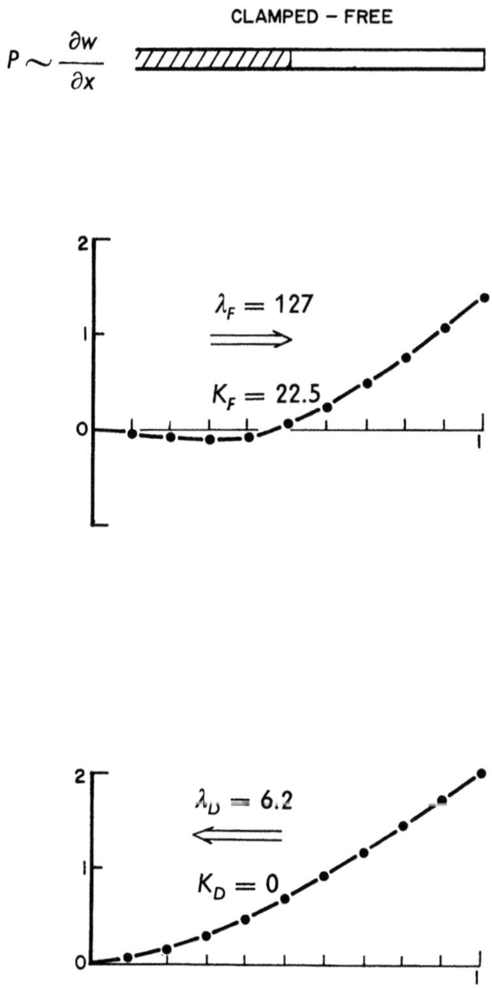

Fig. AII.4

References

[1] Fung, Y. C., *A Summary of the Theories and Experiments on Panel Flutter*. Guggenheim Aeronautical Laboratories, California Institute of Technology, Pasadena, California, *AFOSR TN 60–224* (May 1960). Also AGARD Manual on *Aeroelasticity*, Part III, Chapter 7 (1961).

[2] Bolotin, V. V., *Nonconservative Problems of the Theory of Elastic Stability*. MacMillan Co., New York (1963) 274–312.

[3] Bisplinghoff, R. L. and H. Ashley, *Principles of Aeroelasticity*. John Wiley and Sons, Inc., New York (1962) 416–449.

[4] Johns, D. J., A Survey on Panel Flutter. Advisory Group for Aeronautical Research and Development, *Advisory Report 1* (November 1965).

[5] Johns, D. J., A Panel Flutter Review, AGARD Manual on *Aeroelasticity*, Part III, Chapter 7 (May 1969).

[6] Dowell, E. H., *Nonlinear Theory of the Flutter of Plates and Shells*, Fluid-Solid Interaction Symposium, edited by J. E. Greenspon, American Society of Mechanical Engineers Winter Meeting, Pittsburgh, Pennsylvania (November 1967).

[7] Dowell, E. H., Panel Flutter: A Review of the Aeroelastic Stability of Plates and Shells. *AIAA Journal, 8, 3* (March 1970) 385–399.

[8] Panel Flutter, NASA Space Vehicle Design Criteria, *NASA SP-8004* (1972).

[9] Lemley, C. E., Design Criteria for the Prediction of Panel Flutter. Air Force Flight Dynamics Laboratory, *TR 67–140, I, II* (August 1968).

[10] Dowell, E. H. and H. M. Voss, Experimental and Theoretical Panel Flutter Studies in the Mach Number Range 1.0 to 5.0. U. S. Air Force, Aeronautical Systems Division, *TDR-63–449* (December 1963). Also *AIAA Journal, 3, 12* (December 1965) 2292–2304.

[11] Hess, R. W., Experimental and Analytical Investigation of the Flutter of Flat Built-up Panels Under Streamwise In-Plane Load. *NASA TR R-330* (February 1970).

[12] Muhlstein, L., Jr., P. A. Gaspers, Jr. and D. W. Riddle, An Experimental Study of the Influence of the Turbulent Boundary Layer on Panel Flutter. *NASA TN D-4486* (1968).

[13] Gaspers, P. A., Jr., L. Muhlstein, Jr. and D. N. Petroff, Further Experimental

Results on the Influence of the Turbulent Boundary Layer on Panel Flutter. *NASA TN D-5798* (May 1970).

[14] Maestrello, L., Measurement of Noise Radiated by Boundary Layer-Excited Panels. *Journal of Sound and Vibration, 2,* (1965) 100.

[15] Maestrello, L., Measurement and Analysis of the Response Fields of Turbulent Boundary Layer-Excited Panels. *Journal of Sound and Vibration, 2* (1965) 270.

[16] Maestrello, L., Radiation From and Panel Response to a Supersonic Turbulent Boundary Layer. Boeing Scientific Laboratories Document *Dl-82-0719* (1968).

[17] Gislason, T., Jr., An Experimental Investigation of Panel Divergence at Subsonic Speeds. *AIAA Journal, 9, 11* (Nov. 1971) 2252–2258.

[18] Kappus, H. P., E. C. Lemley and N. H. Zimmerman, An Experimental Investigation of High Amplitude Panel Flutter. *NASA CR-1837* (May 1971).

[19] Muhlstein, L., Jr., A Forced-Vibration Technique for Investigation of Panel Flutter. *AIAA Paper 66–769*, Los Angeles, California (1966).

[20] Dowell, E. H., Nonlinear Flutter of Curved Plates. *AIAA Journal, 7, 3* (March 1969) 424–431.

[21] Dowell, E. H., Fatigue Life Estimation of Fluttering Panels. *AIAA Journal, 8, 10* (October 1970) 1879–1881.

[22] Lock, M. H. and Y. C. Fung, Comparative Experimental and Theoretical Studies of the Flutter of Flat Panels in a Low Supersonic Flow. Guggenheim Aeronautical Laboratories, California Institute of Technology, Pasadena, California, *AFOSR TN 670* (May 1961). Also M. H. Lock, Supersonic Flutter of Two-Dimensional Flat Panels: An Experimental Evaluation of the Theory. *Proceedings of the Fourth U. S. National Congress of Applied Mechanics, 2* (1962) 1351–1357.

[23] Shideler, J. L., S. C. Dixon and C. P. Shore, Flutter at Mach 3 Thermally Stressed Panels and Comparison with Theory for Panels with Edge Rotational Restraint. *NASA TN D-3498* (August 1966).

[24] Dugundji, J. and J. M. Calligeros, Similarity Laws for Aero-thermoelastic Testing. *Journal of the Aerospace Sciences, 29, 1* (August 1962) 935–950.

[25] Kordes, Eldon E. and R. B. Noll, Flight Flutter Results for Flat Rectangular Panels. *NASA TN D-1058* (1961).

[26] Anon., Saturn V Launch Vehicle Flight Evaluation Report *AF-503,* Apollo 6 Mission, Prepared by Saturn V Flight Evaluation Working Group, *MPR-SAT-FE-68–3* (June 25, 1968).

[27] Lin, Y. K., *Probabilistic Structural Dynamics.* McGraw-Hill, New York (1967).

[28] Hedgepeth, J. M., Flutter of Rectangular Simply-Supported Panels at High Supersonic Speeds. *Journal of the Aeronautical Sciences, 24, 8* (August 1957) 563–573.

[29] Cunningham, H. J., Flutter Analysis of Flat Rectangular Panels Based on Three-Dimensional Supersonic Unsteady Potential Flow. *NASA TR-R-256* (1967).

[30] Olson, M. D., Finite Elements Applied to Panel Flutter, *AIAA Journal, 5, 12* (December 1967) 2267–2270.

[31] Dugundji, J., Theoretical Considerations of Panel Flutter at High Supersonic Mach Numbers. *AIAA Journal, 4, 7* (July 1966) 1257–1266.

[32] Yates, J. E., A Study of Panel Flutter with the Exact Method of Zeydel. *NASA CR-1721* (December 1970).

[33] Spriggs, J. H., A. F. Messiter and W. J. Anderson, Membrane Flutter Paradox – An Explanation by Singular Perturbation Methods. *AIAA Journal, 7, 9* (September 1969) 1704–1709.

[34] Dowell, E. H. and C. S. Ventres, On the Flutter of Low Aspect Ratio Plates. *AIAA Journal, 8, 6* (June 1970) 1162–1164.

[35] Miles, J. W., On the Aerodynamic Stability of Thin Panels. *Journal of the Aeronautical Sciences, 23, 8* (August 1956) 771–780.

[36] Dowell, E. H., Flutter of Infinitely Long Plates and Shells. Part I: Plate. *AIAA Journal, 4, 8* (August 1966) 1370–1377.

[37] Dowell, E. H., Theoretical Vibration and Flutter Studies of Point Supported Plates. *J. of Spacecraft and Rockets, 10, 6* (June 1973) 389–395.

[38] Erickson, L. L., Supersonic Flutter of Flat Rectangular Orthotropic Panels Elastically Restrained Against Edge Rotation. *NASA TN D-3500* (August 1966).

[39] Erickson, L. L., Supersonic Flutter of Sandwich Panels: Effects of Face Sheet Bending Stiffness, Rotary Inertia, and Orthotropic Core Shear Stiffness. *NASA TN D-6427* (July 1971).

[40] Gaspers, P. A., Jr. and B. Redd, A Theoretical Analysis of the Flutter of Orthotropic Panels Exposed to a High Supersonic Stream of Arbitrary Direction. *NASA TN D-3551* (August 1966).

[41] Bohon, H. L. and M. S. Anderson, The Role of Boundary Conditions on Flutter of Orthotropic Panels. *AIAA Journal, 4, 7* (July 1966) 1241–1248.

[42] Shore, C. P., Effects of Structural Damping on Flutter of Stressed Panels. *NASA TN D-4990* (1969).

[43] Voss, H. M., The Effect of an External Supersonic Flow on the Vibration Characteristics of Thin Cylindrical Shells. *Journal of the Aerospace Sciences, 28, 12* (December 1961) 945–956.

[44] Olson, M. D. and Y. C. Fung, Supersonic Flutter of Circular Cylindrical Shells Subjected to Internal Pressure and Axial Compression. *AIAA Journal, 4, 5* (May 1966) 858–864.

[45] Dixon, S. C. and M. L. Hudson, Flutter, Vibration, and Buckling of Truncated Orthotropic Conical Shells with Generalized Elastic Edge Restraint. *NASA TN D-5759* (1970).

[46] LeClerc, J., Study of the Flutter of Cylindrical Thin Shells with the Limitations of the Linearized Potential Theory, Parts I and II. *Journal de Mecanique, 9, 1* (March 1970) 111–154 and (June 1970) 213–265.

[47] Parthan, S. and D. J. Johns, Aerodynamic Generalized Forces for Supersonic Shell Flutter. *AIAA Journal, 10, 10* (October 1972) 1369–1371.

[48] Dowell, E. H., Nonlinear Flutter of Curved Plates II. *AIAA Journal, 8, 2* (February 1970) 261–263. Also see comments of Matsuzaki, Y., Natural Vibration and Flutter of Cylindrically Curved Panels. *AIAA Journal, 11, 6* (June 1973) 771–772.

[49] Fralich, R. W., Postbuckling Effects on the Flutter of Simply-Supported Rectangular Panels at Supersonic Speeds. *NASA TN D-1615* (March 1963).

[50] Carter, L. L. and R. O. Stearman, Some Aspects of Cylindrical Shell Panel Flutter. *AIAA Journal, 6, 1* (January 1968) 37–43.

[51] Barr, G. W. and R. O. Stearman, Aeroelastic Stability Characteristics of Cylindrical

Shells Considering Imperfections and Edge Constraint. *AIAA Journal, 7, 5* (May 1969) 912–919.

[52] Kobayashi, Shigeo, Flutter of Simply-Supported Rectangular Panels in Supersonic Flow – Two-Dimensional Panel Flutter, I-Simply-Supported Panel, II-Clamped Panel. Transactions of Japan Society of *Aeronautical and Space Sciences, 5* (1962) 70–118.

[53] Dowell, E. H., Nonlinear Oscillations of a Fluttering Plate I. *AIAA Journal, 4, 7* (July 1966) 1267–1275. Also Part II, *AIAA Journal, 5, 10* (October 1967) 1856–1862.

[54] Evensen, D. A. and M. D. Olson, Nonlinear Flutter of a Circular Cylindrical Shell in Supersonic Flow. *NASA TN D-4265* (1967).

[55] Morino, L., C. C. Kuo and J. Dugundji, Perturbation and Harmonic Balance Methods for Nonlinear Panel Flutter. *AIAA Journal, 10, 11* (November 1972) 1479–1484.

[56] Dowell, E. H., Generalized Aerodynamic Forces on a Flexible Plate Undergoing Transient Motion. *Quarterly of Applied Mathematics, 24, 4* (December 1967) 2267–2270.

[57] Dowell, E. H., Generalized Aerodynamic Forces on a Flexible Cylindrical Shell Undergoing Transient Motion. *Quarterly of Applied Mathematics, 26, 3* (October 1968) 343–353.

[58] Dowell, E. H., Generalized Aerodynamic Forces on a Plate Undergoing Transient Motion in a Shear Flow with an Application to Panel Flutter. *AIAA Journal, 9, 5* (May 1971) 834–841. Also see, Aerodynamic Boundary Layer Effects on Flutter and Damping of Plates. *Journal of Aircraft, 10, 12* (December 1973) 734–738.

[59] Bailie, J. A. and J. E. McFeely, Panel Flutter in Hypersonic Flow. *AIAA Journal, 6, 2* (February 1968) 332–337.

[60] Lyon, R. H. and G. Maidanik, Power Flow Between Linearly Coupled Oscillators. *Journal of the Acoustical Society of America, 34* (1962) 623.

[61] Dowell, E. H. and H. M. Voss, The Effect of a Cavity on Panel Vibration. *AIAA Journal, 1* (1963) 476–477.

[62] Pretlove, A. J., Forced Vibrations of a Rectangular Panel Backed by a Closed Rectangular Cavity. *Journal of Sound and Vibration, 3* (1966) 252.

[63] Strawderman, W. A., The Acoustic Field in a Closed Space Behind a Rectangular Simply-Supported Plate Excited by Boundary Layer Turbulence. U. S. Navy Underwater Sound Laboratory Report No. 827, Fort Trumball, New London, Connecticut, 1967. Also see, Turbulence Induced Vibrations: An Evaluation of Finite and Infinite Plate Models, *Journal of The Acoustical Society of America, 46,* N.S. (November 1969) 1294–1307.

[64] Strawderman, W. A. and R. A. Christman, Turbulence-Induced Plate Vibrations: Some Effects of Fluid Loading on Finite and Infinite Plates. *Journal of the Acoustical Society of America, 52, 5* (November 1972) 1537–1552.

[65] Bies, D. A., A Review of Flight and Wind Tunnel Measurements of Boundary Layer Pressure Fluctuations and Induced Structural Response. *NASA CR-626* (1966).

[66] Dowell, E. H., Transmission of Noise From a Turbulent Boundary Layer Through a Flexible Plate Into a Closed Cavity. *Journal of the Acoustical Society of America, 46* (1969) 348.

[67] Dowell, E. H., Noise or Flutter or Both? *Journal of Sound and Vibration, 11, 2* (1970) 159–180.

[68] Shinozuka, M., Simulation of Multivariate and Multidimensional Random Processes. *Journal of the Acoustical Society of America, 49* (1971) 357–367.

[69] Vaicaitis, R., C. M. Jan, and M. Shinozuka, Nonlinear Panel Response from a Turbulent Boundary Layer. *AIAA Journal, 10, 7* (July 1972) 895–899. Also see Vaicaitis, R., E. H. Dowell and C. S. Ventres, Nonlinear Panel Response by a Monte Carlo Approach. *AIAA Journal, 12, 5* (May 1974) 685–691.

[70] Dowell, E. H., Nonlinear Theory of Unstable Plane Poiseuille Flow. *Journal of Fluid Mechanics, 38* (1969) 401.

[71] George, W. D. and J. D. Hellums, Hydrodynamic Stability in Plane Poiseuille Flow with Finite Amplitude Disturbances. *Journal of Fluid Mechanics, 51*, Part *4* (1972) 687–704.

[72] Dixon, S. C., Comparison of Panel Flutter Results from Approximate Aerodynamic Theory with Results from Exact Inviscid Theory and Experiment. *NASA TN D-3649* (1966).

[73] Ventres, C. S. and E. H. Dowell, Comparison of Theory and Experiment for Nonlinear Flutter of Loaded Plates. *AIAA Journal, 8, 11* (November 1970) 2022–2030.

[74] Dowell, E. H., Flutter of Buckled Plates at Zero Dynamic Pressure. *AIAA Journal, 8, 3* (March 1970) 583–584.

[75] Shore, C. P., Experimental Investigation of Flutter at Mach 3 of Rotationally Restrained Panels and Comparison with Theory. *NASA TN D-5508* (1969).

[76] Shore, C. P., Flutter Design Charts for Biaxially Loaded Isotropic Panels. *Journal of Aircraft, 7, 4* (July–August 1970) 325–329.

[77] Ventres, C. S., Flutter of a Buckled Plate Exposed to a Static Pressure Differential. *AIAA Journal, 9, 5* (May 1971) 958–960.

[78] Anderson, W. J., Experiments on the Flutter of Flat and Slightly Curved Panels at M = 2.18. Guggenheim Aeronautical Laboratories, California Institute of Technology, Pasadena, California, *AFOSR TN 2996* (1962).

[79] Stearman, R. O., M. H. Lock and Y. C. Fung, Ames Tests on the Flutter of Cylindrical Shells. *Structural Dynamics Report SM 63–37*, Graduate Aeronautical Laboratory, California Institute of Technology, Pasadena, California (1962).

[80] Olson, M. D. and Y. C. Fung, Comparing Theory and Experiment for the Supersonic Flutter of Cylindrical Shells. *AIAA Journal, 5, 19* (October 1967) 1849–1855.

[81] Stearman, R. O., An Experimental Study on the Aeroelastic Stability of Thin Cylindrical Shells at the Lower Supersonic Mach Numbers. *AFOSR 66–2828* and *ARL 67–0006*, Midwest Research Institute, Kansas City, Missouri (1966).

[82] Matsuzaki, Y. and S. Kobayashi, Unsteady Supersonic Aerodynamic Forces on an Oscillating Circular Cylindrical Shell. *AIAA Journal, 9, 12* (December 1971) 2358–2362.

[83] Miles, J. W., Supersonic Flutter of a Cylindrical Shell, Part I. *Journal of the Aeronautical Sciences, 24, 2* (September 1966) 107–118.

[84] Dowell, E. H., Flutter of Infinitely Long Plates and Shells Part II: *Cylindrical Shell. AIAA Journal, 4, 9* (September 1966) 1510–1518.

[85] Matsuzaki, Y. and S. Kobayashi, A Theoretical and Experimental Study of Super-

sonic Panel Flutter of Circular Cylindrical Shells. *Proceedings of the Eighth International Symposium on Space Technology and Science*, Tokyo (1969) 281–290.

[86] Paidoussis, M. P. and J. P. Denise, Flutter of Thin Cylindrical Shells Conveying Fluid. *Journal of Sound and Vibration, 20, 1* (1972) 9–26.

[87] Bohon, H. L., M. S. Anderson and W. L. Heard, Jr., Flutter Design of Stiffened-Skin Panels for Hypersonic Aircraft. *NASA TN D-5555*, December (1969).

[88] Golden, C. T., T. R. Hager and L. L. Sherman, Orthotropic Panel Flutter Analysis Correlation, X-20. *The Boeing Company, Document No. D2-81301, AD 446–584* (1964).

[89] Shyprykovich, P. and J. W. Sawyer, Orthotropic Panel Flutter at Arbitrary Yaw Angles – Experiment and Correlation with Theory. *AIAA Paper No. 73–192* (January 1973).

[90] Kobett, D. R., Flutter of Multiple, Streamwise Bay Rectangular Panels at Low Supersonic Mach Number. *NASA CR-538*, Midwest Research Institute (August 1966).

[91] Lock, M. H. and E. F. Farkas, Flutter of Two-bay Flat Panels of Infinite Span at Supersonic Mach Number. *AIAA Journal, 3, 9* (1965) 1692–1697.

[92] Dowell, E. H., Flutter of Multibay Panels at High Supersonic Speeds. *AIAA Journal, 2, 10* (October 1964) 1805–1811.

[93] Dugundji, J., E. Dowell and B. Perkin, Subsonic Flutter of Panels on a Continuous Elastic Foundation. *AIAA Journal, 1, 5* (May 1963) 1146–1154.

[94] Srinivasan, A. V., Flutter Analysis of Rotating Cylindrical Shells Immersed in a Circular Helical Flow Field of Air. *AIAA Journal, 9, 3* (March 1971) 394–400.

[95] Dowell, E. H. and A. V. Srinivasan, Aeroelastic Stability of Cylindrical Shells Subjected to a Rotating Flow. *AIAA Paper No. 74–142*, presented at AIAA 12th Aerospace Sciences Meeting (January 1974).

[96] Ventres, C. S. and C. K. Kang, Large Amplitude Flutter of a Low Aspect Ratio Panel at Low Supersonic Speeds – Comparison of Theory and Experiment. *Princeton University AMS Report No. 1116* (August 1973).

[97] Maestrello, L., Radiation from and Panel Response to a Supersonic Turbulent Boundary Layer. *Journal of Sound and Vibration, 10, 2* (1969) 261–295.

[98] Wilby, J. F., The Response of Simple Panels to Turbulent Boundary Layer Excitation. *AFFDL-TR-67–70* (Oct. 1967).

[99] Chyu, W. J. and M. K. Au-Yang, Random Response of Rectangular Panels to the Pressure Field Beneath a Turbulent Boundary Layer in Subsonic Flows. *NASA TN D-6970*, Ames Research Center (October 1972).

[100] Chyu, W. J. and M. K. Au-Yang, Response of Panels to Turbulence – Induced Surface-Pressure Fluctuations and Resulting Acoustic Radiation to the Flow Field. *AIAA Paper 73–993*, AIAA Acoustic Specialists Conference, Seattle (October 1973).

[101] Volmir, A. S., A Translation of Flexible Plates and Shells. Air Force Flight Dynamics Laboratory, Wright-Patterson Air Force Base, Ohio, *AFFDL TR 66–126* (1967).

[102] Washizu, K., *Variational Methods in Elasticity and Plasticity*. Pergamon Press (1968).

Subject index

Subject index

Author index

9 789028 604049

GPSR Compliance

The European Union's (EU) General Product Safety Regulation (GPSR) is a set of rules that requires consumer products to be safe and our obligations to ensure this.

If you have any concerns about our products, you can contact us on ProductSafety@springernature.com

In case Publisher is established outside the EU, the EU authorized representative is:

Springer Nature Customer Service Center GmbH
Europaplatz 3
69115 Heidelberg, Germany

Batch number: 10584448

Printed by Printforce, the Netherlands